T0258895

ROUTLEDGE LIBRARY EDITIONS: FORESTRY

Volume 1

PLANNED MANAGEMENT OF FORESTS

PLANNED
MANAGEMENT
OF FORESTS

N. V. BRASNETT

Routledge
Taylor & Francis Group

LONDON AND NEW YORK

First published in 1953 by George Allen & Unwin Ltd.

This edition first published in 2024
by Routledge
4 Park Square, Milton Park, Abingdon, Oxon OX14 4RN

and by Routledge
605 Third Avenue, New York, NY 10158

Routledge is an imprint of the Taylor & Francis Group, an informa business

© 1953 N. V. Brasnett

British Library Cataloguing in Publication Data
A catalogue record for this book is available from the British Library

ISBN: 978-1-032-77116-8 (Set)
ISBN: 978-1-032-76682-9 (Volume 1) (hbk)
ISBN: 978-1-032-76689-8 (Volume 1) (pbk)
ISBN: 978-1-003-47961-1 (Volume 1) (ebk)

DOI: 10.4324/9781003479611

Publisher's Note
The publisher has gone to great lengths to ensure the quality of this reprint but points out that some imperfections in the original copies may be apparent.

Disclaimer
The publisher has made every effort to trace copyright holders and would welcome correspondence from those they have been unable to trace.

PLANNED
MANAGEMENT
of
FORESTS

N. V. BRASNETT

M.A., DIP.FOR. (*Cantab*)

*Lecturer in Forest Management in the
Imperial Forestry Institute, Oxford,
formerly of the Colonial Forest Service*

George Allen & Unwin Ltd

RUSKIN HOUSE MUSEUM STREET LONDON

PRINTED IN GREAT BRITAIN
in 11-point Baskerville type
BY SIMSON SHAND LTD
LONDON AND HERTFORD

PREFACE

This book, which contains nothing original, has been compiled to provide students of forestry with a simple outline of what the management of forests involves, and of the ways in which foresters, working in various conditions, have attempted to organize and control their operations.

I am indebted to many authors and practising foresters for what I have set down, but above all to one of my predecessors (1922–37) as lecturer in forest management at Oxford, the late Mr. Ray Bourne. Not only have many of his teachings become standard British practice, particularly in the tropics, but also he was good enough to lend me in my early days at Oxford a manuscript of his own, and to permit me to take notes from it for use in my lectures. These notes were the prototype of Part III of this book, but have been so amended, rearranged, condensed, and added to that only the historical facts remain of what Bourne wrote. Even in respect of these I have sometimes preferred to accept an authority whom he had apparently rejected. Bourne therefore cannot be blamed for any shortcomings, but it was he who drew my attention to the value of such an historical study and provided guidance for it.

Some readers may wonder why this history is placed last instead of first. The reason is a belief that the history is more useful and interesting to a student who already has a grasp of what was being evolved and so can appreciate the significance of the various stages. Also I wished to keep the earlier chapters as simple as possible. This arrangement necessitates a certain amount of division and repetition, for instance, Biolley's *Méthode du Contrôle* is outlined in Chapter 12 and discussed in detail in Chapter 23. Cross-references have been provided and I trust the inconvenience caused will be small.

Finally, I have to thank Professor H. G. Champion for his advice and encouragement.

N. V. BRASNETT

Oxford

CONTENTS

CONTENTS

PART I

THE GROWTH
AND ORGANIZATION
OF
FOREST CROPS

CHAPTER

I

Forest management is not a subject in itself, but is the practical application of science, technology, and economics to a forest estate for the production of certain desired results.

Management of forest areas is a practical activity, a duty which should be entrusted to trained foresters and the main reason why such men should be trained. It consists essentially in assessing the potentialities of a given area, in deciding what within these potentialities is desired from it, in organizing the area to achieve this object to the maximum extent consistent with the wellbeing of the soil, in planning the operations necessary to this end, and in carrying them out from day to day.

The primary object of good management is provision of the maximum benefit to the greatest number of people for all time.

The word forest inevitably brings to mind the idea of timber trees, and forestry suggests the growing and harvesting of trees. Forestry is, however, much more than this. Some of the areas which foresters have to manage have practically no trees on them, and some are quite incapable of growing timber trees. Tracts may be handed over to foresters to manage for what are known as indirect benefits, such as the maintenance of stream flow, prevention of erosion, the protection of agricultural land from floods or wind, and so on. Other areas may require management in the public interest so as to preserve the beauty of natural scenery, to provide facilities for camping, recreation, and hunting, or to increase their stock carrying capacity, which is known as range management in America.

Much of a forester's work has, of course, to do with production, that is to say the organization and growing of steady supplies of wood or other products of the forest, such as resin, tannin bark, or gum. Such work may aim at ensuring the supply of ample wood for farm buildings and implements, fencing, cooking, and heating, or at providing the raw material to keep sawmills, plywood, or paper mills working.

The detailed objects of management, which may, of course, include more than one of the intentions mentioned above, must be decided for each individual forest or group of forests. The area must then be organized for the attainment of the objects and a plan prepared and executed so that the basic assets, which are the soil and climate, will be used to their maximum capacity in that attainment.

Forest management must be founded on the sciences and skills of geology, pedology, botany, ecology, silviculture, and economics in the selection and treatment of vegetation, and of engineering and marketing in the harvesting, extraction, and preparation of crops. Any association of plants is a living entity, the result of the interaction of soil, climate, surrounding vegetation, and of any interference by man or other animals, as by cutting, burning, grazing, and so on. Vegetation is not static but changes gradually, as for instance when one association of species produces conditions favourable for the germination and development of the young of other species. A forester must understand what is happening in each part of a forest before he can devise the best way to manage it. A fixed pattern of management cannot be applied to a forest as it can to a collection of machinery in a factory, nor, because a certain forest has been managed in one way with successful results, does it follow that another forest, though apparently similar, will respond in the same way to the same system. Ready-made systems of management cannot be taken off a peg and applied to various classes of forest. There are local manuals of the management methods normally used in restricted localities, and there are technical schools which teach the standard procedures of such areas, but university education aims at something wider than this, the basic understanding of the processes of nature and how advantage may be taken of them to satisfy the selective demands of man.

The basis of management must be an understanding of an individual forest based on skilled observation and scientific deduction from the facts observed. For this the forester needs all the knowledge of the basic sciences he can obtain, but even more than that he needs experience, which can only be gained by missing no opportunity to practise observation and deduction, and to check his reading of a forest whenever possible. A great deal can be learnt, however, from a study of what has been done by foresters in various circumstances, the systems of management they have devised for various sets of conditions, the reasons why these were adopted, and the results which have followed their application.

The most complicated and generally the most common of the problems of forest management are concerned with forests which are maintained for the production of timber. Such production requires the investment of capital in soil for the raising of timber crops just as farming needs capital in the form of soil for the raising of food crops. Most food crops such as wheat are annual, and the capital of the soil plus that invested in ploughing, sowing, and harvesting gives an annual return, or interest, in the form of a crop. In order to produce an annual income a forest requires an additional form of capital, a capital of trees on the soil, upon which the annual accretions of new wood can be laid each year. The year's growth cannot be peeled off each tree, and it is impossible to say by looking at a forest at the end of the growing season what is the year's growth and where it can be harvested. Some of each tree is capital and some is interest.

How much wood can be cut without reducing the capital?

Has the right amount of capital been invested on the soil to produce the maximum rate of interest from that soil in the existing climate?

Should some of the interest be left in the forest to increase the capital, or should more than the interest be cut out because the forest is over-capitalized?

These are some of the problems of management, quite apart from the silvicultural ones of whether the species most suited to the soil are being grown in the best way, and the economic ones of whether the production satisfies the market and is that for which the best price will be paid.

In order to keep a sawmill or any other market which depends on a forest constantly supplied, a steady flow of produce has to be made available. Trees should only be cut in accordance with the principles of good silviculture, but to achieve this steady supply, which is called a sustained yield, a forecast has to be made of the amount which it should be possible to cut annually or over a period of years from a forest without detriment to its productivity. A useful starting point for the consideration of forest capital or GROWING STOCK and its interest or INCREMENT is the theoretical conception of a NORMAL FOREST. This conception, which embodies certain fallacies to be discussed later, was the basis of many of the earlier systems of management and yield regulation.

A normal forest itself is not a theoretical conception, though it is very rarely attained completely, and its most frequent use is that of a standard with which to compare an actual forest to bring out the deficiencies of the latter for sustained yield management. A normal forest is an ideally constituted forest with such volumes of trees of various ages so distributed and growing in such a way that they produce equal annual volumes of produce which can be removed continuously without detriment to future production.

The simplest example to consider is that of a firewood plantation in the tropics consisting of ten acres of Eucalypt trees, one acre of which has been planted each year for ten years. There are therefore ten age-gradations of equal size constituting a NORMAL SERIES OF AGE GRADATIONS of which one is available for clearfelling each year. At the end of the tenth year the plot planted first is cut and allowed to coppice. At the end of the next year this regenerated plot has become the one-year old age-gradation, the series is complete again and the oldest plot, now ten years old, is felled. This is illustrated diagrammatically in Fig. 1, where the age gradation areas are shown along the base AB with the theoretical volume standing on each, assuming for the moment that an acre of plantation lays on an equal volume of wood in each year of its life. In the diagram the ten-year old acre has grown 5,000 cu. ft., so the MEAN ANNUAL INCREMENT at that age is $\frac{5,000}{10} = 500$ cu. ft., and this is shown as the volume on the one-year old acre as i. The volume on the two-year old acre is shown as $2i$ and so on up to the volume of the ten-year old acre as $r \times i$, in which r stands for the number of years which has been chosen to elapse between the time of planting and final felling, called the ROTATION.

This $r \times i$ which is the volume standing on the oldest age-gradation at the end of r years (in this case $10 \times 500 = 5,000$) is also the sum of the mean annual increments of all the r (10) age-gradations, and may be called I to represent the increment of the whole series. Therefore by felling the rotation-age acre each year the NORMAL INCREMENT for that rotation of this Normal Series is being harvested.

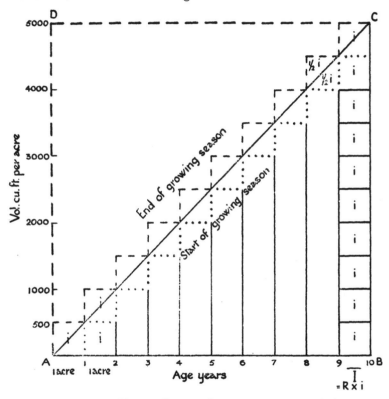

FIG. I THEORETICAL NORMAL GROWING STOCK ON A SERIES OF R (10) EQUAL (I ACRE) AGE GRADATIONS.

Based on the assumption that each acre puts on an equal annual increment of i (500) cubic feet in each year of the rotation of R (10) years.

Volume of Oldest Age Gradation at end of R years $= R \times i = I =$ Mean Annual Increment of whole series (5,000 cubic feet).

Normal Growing Stock at Middle of Growing Season on R acres $= I \times \frac{1}{2}R = 25,000$ cubic feet.

Normal Growing Stock at End of Growing Season on R acres $= I \times \frac{1}{2}R + \frac{1}{2}I = 27,500$ cubic feet.

Normal Growing Stock at Start of Growing Season on R acres $= I \times \frac{1}{2}R - \frac{1}{2}I = 22,500$ cubic feet.

The Normal Growing Stock of the series at the end of the growing season in the rotation-age year (the tenth in this case) is the sum of the rectangles standing on the r (10) age-gradations, which can be calculated

$$i + 2i + 3i + \ldots + ri = (i + ir) \times \frac{r}{2} = (i \times r)\frac{r}{2} + \frac{i \times r}{2}$$

$$= (I \times \frac{r}{2}) + \frac{I}{2}$$

$$500 + 1{,}000 + 1{,}500 + \ldots + 5{,}000 = (5{,}000 \times \frac{10}{2})$$

$$+ \frac{5{,}000}{2} = 27{,}500$$

Similarly before the rotation age growing season there will be one less Normal Annual Increment for the whole series standing on the ground and the Normal Growing Stock

$$= (I \times \frac{r}{2}) - \frac{I}{2}.$$

In the middle of the growing season therefore N.G.S.

$$= I \times \frac{r}{2} \quad [5{,}000 \times 5 = 25{,}000.]$$

This last formula was the way in which this method of calculating a normal growing stock was generally used, and it should be noted that it gives the stock on r acres, so the average N.G.S. per acre in our example would be 2,500 cu. ft.

All this may be seen graphically on the diagram where the line AC includes below it half of the tenth year's increment on each of the age gradations, and the area of the triangle ABC (the perpendicular I multiplied by half the base r) represents the volume of the N.G.S. in the middle of the tenth year's growing season. The ten half squares above and below the line AC representing half the mean annual increment on each age gradation may be added to or subtracted from the area of the triangle to give the N.G.S. after and before the growing season respectively.

Before leaving this diagram it may be pointed out that by substituting AGE CLASSES of ten years each for the age gradations it can represent a normal series of (say) pine stands 1–10, 11–20, 21–30, etc., years old for a 100 year rotation. An age class is merely a group of age gradations, say ten, twenty, or thirty, and is more often used than the latter because of the small growth of a tree in one year.

The sum of a series of age gradations or classes in a forest is thus the capital or growing stock of the forest, and the sum of a normal series is the normal growing stock. As we shall see later it is not necessary for the age classes to be on separate areas for a forest to be normal. The ages can be all mixed up, seedlings under mature trees, etc., as long as the proportions are right and a normal increment for the rotation is being laid on. Actually the proportions of area occupied by the age classes in an uneven-aged forest are not quite the same as in one composed of even-aged stands, because the space can be shared by small trees growing under big ones.[1]

The growing stock of a forest may have the correct volume for its normal growing stock, but if it is not properly distributed among the age classes so that an annual or periodic felling of trees of rotation age can be made so as to result in equal annual or periodic yields, the forest will not be normal. For instance, five nine-year old acres and five two-year old acres of the Eucalypts shown in the diagram would have an end of season growing stock of 27,500 cu. ft., but this forest would not be normal because you could not possibly fell equal annual or periodic yields of ten-year old trees.

Similarly, the annual increment of a forest may be the correct volume for a normal increment, but unless it is laid on to trees of the right size classes in the right proportions it is not a normal increment.

Furthermore, a forest can only be normal for one particular length of rotation. If in our example the rotation of ten years is changed to twelve years, there would not be a normal series of age gradations in such proportions that the felling of the oldest at twelve years would result in a constant equal annual yield. The series would have to be adjusted, for instance, by stopping felling for two years and planting an extra acre in each of these years, to make it normal for a twelve-year rotation. Conversely, if the rotation was dropped to eight years the surplus nine- and ten-year old plots would have to be cut gradually until there were one and a quarter acres of each age gradation to be felled at eight years old each

[1] See pages 213, 219.

year. This illustrates an unalterable principle of forest management, that if you lengthen a rotation you have to increase your growing stock, and in doing so face a temporary drop in revenue, whereas, if you decrease your rotation, you have surplus growing stock to dispose of.

It has been stated that Fig. 1 represents a theoretical conception of a normal forest because it shows equal annual increments put on by the trees of each acre throughout their life. Actually the volume of timber put on by very small trees before they reach timber size is nil. Their annual increment increases rapidly in middle age, and then slows down again as the trees grow old. The actual increase in volume each year is known as the CURRENT ANNUAL INCREMENT to distinguish it from the mean annual increment, which is the volume at any age divided by that age. It is not usually possible to measure one year's growth, so the growth is usually measured for short periods of, say, five years, and the average annual growth over this period, which is really the periodic mean annual increment, is taken as the current annual increment for the last year of the period.

The usual relationship between M.A.I. and C.A.I. is shown in Fig. 2. As the M.A.I. is simply the average of the C.A.I.s to any age, it is obvious that as long as the C.A.I. is greater than the M.A.I. the latter must continue to rise, even if the C.A.I. is falling. When the two are equal the M.A.I. is stationary, and when the C.A.I. is less than the M.A.I. the latter is falling.

An example will illustrate this:

If Vol. at 10 years is 1,000 cu. ft., then M.A.I. at 10 years

$$= \frac{1,000}{10} = 100 \text{ cu. ft.}$$

If C.A.I. in 11th year is 100 cu. ft., then M.A.I. at 11 years

$$= \frac{1,100}{11} = 100 \text{ cu. ft.}$$

If C.A.I. in 11th year is 111 cu. ft., then M.A.I. at 11 years

$$= \frac{1,111}{11} = 101 \text{ cu. ft.}$$

If C.A.I. in 11th year is 89 cu. ft., then M.A.I. at 11 years

$$= \frac{1,089}{11} = 99 \text{ cu. ft.}$$

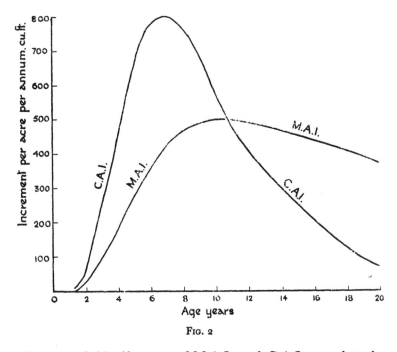

FIG. 2

Because of this, if curves of M.A.I. and C.A.I. are plotted against age as in Fig. 2, the age at which they cross is that at which the M.A.I. has attained its maximum. If the stand is felled at that age the greatest possible volume per acre per annum will be obtained. In Fig. 1 the progress of volume production with age was shown as a straight line, AC, depicting the M.A.I. of the final rotation-age volume. If the actual volumes produced at each age are plotted over age the result is a curve such as that shown in Fig. 3.

In this figure the real normal growing stock for any rotation is represented by the area below the curve to a point on the curve vertically above rotation age. On the diagram are shown also the theoretical normal growing stock triangles according to the equal annual increment conception for rotations of 10, 16, and 20 years. For the 10 year rotation the triangle is the same as that shown in Fig. 1, and is obviously considerably larger than the real N.G.S., as shown by the area under the curve. At 16 years, which is several years after the M.A.I. of this Eucalypt species on the particular site has

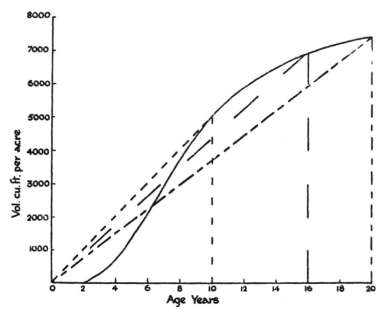

FIG. 3

Rotation years	Growing Stock under curve, cu. ft.	Growing Stock by M.A.I. triangle, cu. ft.	On acres
10	18,310	25,000	10
16	55,245	55,920	16
20	80,740	74,300	20

culminated, the triangle and the area under the curve have approximately the same size. Later the true N.G.S. is greater than the theoretical $I \times \frac{1}{2}r$, as is illustrated at 20 years.[1]

A Swiss forester, named Flury, actually calculated values at various ages for a constant 'c' to substitute for the $\frac{1}{2}$ in the formula N.G.S. $= I \times \frac{1}{2}r$ for use with various species of tree. For instance, for Scots Pine in the conditions of North Germany he found that 'c' would have to be 0·387 for an age of 60 years, and 0·508 for an age of 120 years. These figures

[1] The figures plotted are within the range of the growth of Eucalypts in the tropics and demonstrate exaggerated differences within a short term of years.

were, of course, entirely empirical, and would have to be determined for each species, for each rotation, and each set of growing conditions, and used in the formula N.G.S. $= I \times cr$.

So far one factor which complicates the relationships between yields, and between C.A.I. and M.A.I., has been omitted. The Eucalypts of Fig. 1 were planted at, say, 10 ft. by 10 ft., 435 to the acre, and the end of 10 years about the same number per acre were standing. In this country Scots Pine are often planted at $4\frac{1}{2}$ ft. by $4\frac{1}{2}$ ft., 2,150 per acre, and are thinned out gradually until a final crop of some 150 per acre remains at a hundred years on quality class I sites. The mathematical relationships between C.A.I. and M.A.I. depend on all the measurements being made on the same growing body. In practice it is usual to determine the C.A.I. by laying out a sample plot, thinning it, measuring it now, and then measuring it again after an interval of years before doing another thinning. The thinnings, of course, must be measured, too, so that the total production at any age is known. This, divided by the age, gives the M.A.I. comparable to the C.A.I. measured as above, which is equal to the difference between the total production at any two ages divided by the number of years between measurements.

In some cases records of stands do not include the measurements of past thinnings and then increment calculations have to be based on measurements of the main crop only. It should be remembered that what is the main crop after one thinning grows to become the main crop and the next thinning during the period between measurements. The following examples show in the first three columns the sort of figures per acre which might be obtained from measurements, and in the next four columns information derived from them.

AGE.	MAIN CROP, cu. ft.	THINNINGS, cu. ft.	FINAL CROP, cu. ft.	TOTAL PRODUC- TION, cu. ft.	M.A.I., cu. ft.	C.A.I. per annum.
25	2,400	100	2,500	2,500	$\frac{2,500}{25}$ $= 100$	
						$\frac{3,300-2,500}{5}$ $= 160$

23

AGE	MAIN CROP, cu. ft.	THINNINGS, cu. ft.	FINAL CROP, cu. ft.	TOTAL PRODUC- TION, cu. ft.	M.A.I., cu. ft.	C.A.I. per annum
30	3,000	200	3,200	3,300	$\dfrac{3,300}{30}$ $= 110$	
						$\dfrac{4,060-3,300}{5}$ $= 152$
35	3,500	260	3,760	4,060	$\dfrac{4,060}{35}$ $= 116$	

It should be noted that at 25 years there are 2,500 cu. ft. standing on a sample acre, and that this would be the yield obtained by clear-felling at that age. A thinning of 100 cu. ft. is made and the 2,400 remaining grow into 3,200 cu. ft. by 30 years, an increment of 800 cu. ft. in 5 years, a periodic mean annual increment or C.A.I. for the 5-year period of 160 cu. ft. per acre per annum. Thinnings are intermediate yields, and, from the time they are measurable or saleable, constitute part of the volumetric and financial yield of a forest. Calculations of possible or legitimate yields are sometimes made for final crop trees only, in which case the result is called a FINAL YIELD. When this is done an estimate is made of the volume and value of thinnings which are likely to be made annually on silvicultural grounds, and if a forest is fairly normal it is assumed that these will be approximately equal each year. Sometimes a TOTAL YIELD is calculated, some of which will be harvested as final crop trees and some as thinnings.

What is meant by a legitimate yield is a normal yield from a normal forest, or when a forest is not normal, a yield the cutting of which annually for a period of years will bring the forest nearer to a state of normality.

In the case of the theoretical normal forest of Fig. 1 the theoretical N.G.S. $= \dfrac{I \times r}{2}$ whence $I = \dfrac{2\,\text{G.S.}}{r}$. This normal increment is the normal yield and it is obvious from the diagram that during a rotation the growth on all the age gradations would be $I \times r$ (the rectangle ABCD) or twice the

original growing stock. During the rotation the original growing stock (the triangle ABC) plus half the new increment
during the rotation (the triangle ACD) can be cut, and then
the other half of the increment (the reconstituted triangle
ABC) should be left to be the new growing stock.

When yield tables are available the actual normal growing
stock can be calculated for any rotation from them, and the
theoretical normal growing stock based on the M.A.I. at rotation age only can be disregarded.

A yield table gives the final crop volumes per acre at regular
intervals and if these are plotted to scale on a graph the area
below the curve can be measured. With much less trouble the
volume of the normal growing stock can be calculated direct
from the tables provided it is remembered that the final crop
volume shown for each age interval is not the average volume
per acre of all the age gradations since the last age interval,
but the average volume per acre of the age gradations half-
way through one interval to half-way through the next.

Figures from an imaginary yield table for a species growing
as some pines do in South Africa are given below to illustrate
the calculation of the average mean normal growing stock per
acre for a rotation of 35 years.[1]

AGE, years.	FINAL CROP, cu. ft.	Calculation of total Growing Stock and hence Mean Normal Growing Stock per acre		
		2½ acres	0 – 2½ years old carry	0 cu. ft.
5	0	5 ,,	2½– 7½ ,, ,,	0 ,,
10	300	5 ,,	7½–12½ ,, ,,	1,500 ,,
15	1,000	5 ,,	12½–17½ ,, ,,	5,000 ,,
20	1,750	5 ,,	17½–22½ ,, ,,	8,750 ,,
25	2,650	5 ,,	22½–27½ ,, ,,	13,250 ,,
30	3,500	5 ,,	27½–32½ ,, ,,	17,500 ,,
35	4,000	2½ ,,	32½–35 ,, ,,	10,000 ,,

(Quick △   35 acres   0 –35 years old carry   56,000 cu. ft.
Method 13,200 − 2,000   1 acre Mean N.G.S. =   1,600 cu. ft.

$$= \frac{11,200}{7 \text{ age classes}} = 1{,}600 \text{ per acre}).$$

In practice it is unnecessary to multiply each final crop

[1] See also Appendix I.

25

figure by the age interval and to divide the sum by the rotation, and it is simpler to add up the final crop column direct and to divide it by the number of age intervals (or age classes) added, always remembering that the rotation age final crop figure must be halved before it is added in. In the example above the sum of the final crop figures for ages 5 to 35 less half of the figure for 35 is 11,200 and, as 7 age classes have been added, this must be divided by 7 to give the answer 1,600 for one acre.

Fig. 4 illustrates how the calculation of a series of volumes

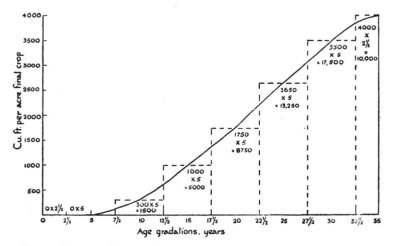

FIG. 4 NORMAL GROWING STOCK FOR A ROTATION OF THIRTY-FIVE YEARS PLOTTED FROM A YIELD TABLE (FICTITIOUS) WITH AN AGE-INTERVAL OF FIVE YEARS.

(Note that the small portions of the measured rectangles above the curve are balanced approximately by the unmeasured portions below the curve.)

for age classes of the interval number of acres is equivalent to measuring rectangles of which the portions measured outside the curve are balanced by unmeasured portions below the curve. This will only be the case when the age interval is relatively short and consequently the series of straight lines substituted for the curve follows the curve fairly closely.

The calculation made above can be expressed in a mathematical formula, N.G.S. on r acres

$$= n(Vn + V2n + V3n + \ldots + Vr-n + \frac{Vr}{2})$$

in which Vn, $V2n$, etc., are the volumes at successive age
intervals

n is the number of years in the age interval

r　　,,　　,,　　,,　　,,　　,,　　rotation

The arithmetical evolution of the formula according to arithmetical progression is given in various textbooks.

The normal growing stock thus calculated is that in the
middle of the growing season.

It must not be overlooked that yield tables give volumes for
unit acres for fully stocked areas, and consequently the normal
growing stock calculated from them will be that for a fully
stocked normal forest, which is practically impossible of
attainment.

CHAPTER

2

ROTATIONS

We have seen that the age at which a forest crop is felled
has a considerable effect upon the volume of wood which can
be produced from a given site. The word ROTATION which
has already been used may be defined as 'The number of
years fixed by the working plan between the formation or regeneration and the final felling of a forest crop'.

When trees are planted and later clear-felled the rotation
is a definite number of years, though it will often happen that
they will not be felled at exactly the age planned. When a crop
of mature trees is felled gradually in order to establish seedlings beneath it over a period of years, and these seedlings in
their turn are felled gradually at maturity, the rotation is the
average life of the trees in the crop, which is equivalent to the
period of years between similar stages in the two harvests, etc.,
the start of regeneration fellings. When this same treatment is
applied to various plots throughout the forest the rotation is
the average life of the crops in the forest. Under the selection
system, when individual trees are cut out of all-aged stands,
the rotation, if it is recognized at all, is the average age at
which trees reach the size it is desired to produce. There are
various types of rotation which may be adopted according to
the objects which it is desired to obtain.

Physical Rotation The rotation which coincides with the
natural lease of life of a tree on a given site. This is only of
importance in protection forests and amenity forests.

Silvicultural Rotation The rotation through which a species
retains maximum vigour of growth and reproduction on a
given site, and at the end of which natural regeneration has
the most favourable conditions. The length of this depends
not only on the effect of age on the quantity and fertility of
the seed produced, but also on the effect of the crowns of the
trees on the soil and vegetation conditions below them at

different ages. Fortunately the upper and lower limits of the silvicultural rotation cover a fairly wide period within which several other kinds of rotation fall.

Technical Rotation The rotation under which a species yields the maximum quantity of a specified size or suitability for economic conversation or special use. The firewood rotation of 10 years in Fig. 1 was a technical rotation, because after 10 years the trees would be too large to cut up into firewood billets without the extra expense of splitting them. A special technical rotation of 250 years is adopted in some State oak forests in France, because logs of the diameter attained at that age split into the maximum number of the largest size of wine-cask staves for their volume with a minimum of waste. It would pay better to cut the trees earlier, but the French government recognizes a duty to the essential wine trade to provide staves for this necessary size of cask.

Rotation of Maximum Volume Production The rotation under which a forest yields the greatest quantity of material per unit area. Its length coincides with the age at which the M.A.I. culminates. Suppose the M.A.I. of a species on a certain site culminates at 100 cu. ft. per acre per annum at the age of 50 years. At 50 years you can cut 5,000 cu. ft. off one acre, repeat this and take 10,000 cu. ft. off the acre in 100 years. After 50 years the M.A.I. falls off ; say it is 75 cu. ft. per acre per annum at the age of 100. Then on a 100-year rotation you would get only 7,500 cu. ft. off one acre in 100 years. Similarly before 50 years old the M.A.I. has not reached its maximum ; suppose it was 75 cu. ft. per acre at the age of 25. Then four times in 100 years you could cut on a rotation of 25 years a final yield of 1,875 cu. ft., 7,500 cu. ft. in all.[1]

Rotation of Highest Income The rotation which yields the highest average annual income, calculated without interest and irrespective of the time when items of revenue and expenditure occur.

Financial or Economic Rotation The rotation which gives the highest net return on capital value, i.e. that under which the soil expectation value calculated with a given rate of interest is at its maximum.

[1] It is assumed that in all these cases the crops are subject to similar thinning regimes. The age of culmination of M.A.I. can be affected by thinning.

The two last-named types of rotation require to be studied against a general background of forest economics.

Before selecting a rotation under which a forest will be worked a forester has to consider the objects of management, the silvicultural requirements of the species, markets, and finance. The rotation selected will generally be a combination of technical and silvicultural rotations, tempered by some economic considerations.

Economic arguments, however, must not be allowed to dictate the adoption of a rotation which might endanger the productivity of the soil. This is the primary capital of a forest, but it does not appear among the mathematical expressions of the formulæ of economists. There may, for instance, be cases in which the frequent exposures of the soil involved in growing crops on a short rotation under a system of clear felling may cause soil deterioration. Certainly the replacement of mixed forests by short-rotation spruce which was adopted for financial reasons in Saxony early in the 19th century gave high yields at first, but within a hundred years the yields had fallen off disastrously. As spruce is a conifer and was planted pure outside its natural habitat the short rotation was not the only cause of the trouble, but it contributed. Local markets for small dimension produce often appear attractive, particularly to private forest owners, but if there are large afforestation schemes nearby, the output from thinnings from these may depress the prices for small sizes. A balanced output of saw timber and of poles and stakes from tending operations will generally be a safer proposition than reliance on one type only.

Once an all-aged forest of species capable of management by selection has been built up there is no longer any question of rotation. Each tree of good shape is allowed to stand for as long as it continues to grow well and is not interfering with more promising trees or the all-aged condition of the stand. The larger such a tree can be allowed to grow the more valuable it will be per cubic foot when it is cut, because large timber is almost always scarcer than small timber and so fetches a better price. This is likely to be even more true in the future than it is to-day. While waiting for the best trees to put on this value increment the owner can cut out his less promising trees at a younger age to bring in his annual revenue.

FOREST ORGANIZATION

Areas which foresters are called upon to manage will, almost certainly, not be normal forests. They may be natural forests which have not been managed before, or, if they have, not for long enough to be approaching normality. They may be artificial plantations not making up a normal series, or a mixture of the two in which some attempt at management has been made. Sometimes forests have to be dealt with which are not large enough to produce annual yields which would be economic to harvest. It may be possible to group such forests into units which are large enough to provide economic sustained yields, but often, owing to diverse ownership or inconvenient location, this is not possible, and individual forests have to be worked for intermittent yields with periods of practically no production between. Sometimes, on the other hand, management has to be started in inaccessible forests for whose products there is very little demand. If a forest cannot be worked it is difficult to give it any silvicultural treatment and so to improve its condition. In such conditions a forester's primary concern is the creation of markets for the potential production of the forest.

The organization of a forest estate, which means the bringing of a forest into regular working in such a way as to fulfil the objects of management agreed upon, forms an important part of forest management. It involves planning for a considerable period, and if the organization is to be carried through without waste of effort, the scheme which has been worked out must be put on paper and adhered to in its main essentials. Such a plan need not be very detailed as regards the distant future, but it must lay down the broad framework into which the operations to be carried out year by year for the next few years are fitted. It must provide for continuity of progress along agreed lines towards an accepted goal, and it must be safeguarded from sabotage by anyone who suddenly thinks that it might be a good idea to do something quite different. This does not mean that the original plan should be immune from amendment: it should, in fact, be revised every few years, but its main provisions should be protected from drastic alteration without careful inquiry into all the circumstances.

Before considering the best way in which such a plan can be drawn up there are certain fundamental matters relating to the form and distribution of forest crops, and the division of the forest into units to facilitate orderly management, which must be considered.

The silvicultural system or systems to be adopted to achieve the desired objects of management will depend on the species and local conditions found in the various parts of the forest. No circumstances can ever justify silviculture which is not based on the species, soil, climate, exposure, aspect, etc., of the area, and sound management can only be based on sound silviculture.

Silvicultural systems fall roughly into two classes: Regular systems, such as clear-felling and shelterwood (uniform), and Irregular systems, such as selection and irregular shelterwood. Regeneration may be obtained by planting, artificial sowing, natural seeding, coppice, suckers, and pollards, in blocks, strips, wedges, groups, and individual spots.

The form of the crops and the distribution of the age classes produced from time to time by the application of various systems vary very considerably. If the felling of mature trees and establishment of regeneration is localized the factor of area can be used by the forest manager, at any rate to some extent, to estimate and regulate the volume output of his forest. If there are trees of all sizes mixed up all over the forest the legitimate yield can only be estimated and regulated in terms of the volume of the whole growing stock and its rates of increment.

The problem of the forest manager is to handle his soil and crops in the most suitable silvicultural manner, but also to organize the silvicultural operations in such a way that convenient quantities of produce of various suitable sizes are available regularly from convenient places, and steady employment is provided for approximately the same number of men each year. Various devices by which this may be facilitated will be described and, when the problems of yield regulation are considered in detail, it will be seen that certain broad distinctions can be drawn between different types of management.

There is, however, no such thing as a definite management system. Certain systems of silviculture, which are now well known, were evolved by foresters, mainly in Europe, to suit the local conditions of particular forests, and have been adopted by others elsewhere, sometimes with success because the conditions were similar, and sometimes without success. A silvicultural system was often combined with a particular form of yield calculation and regulation, and sometimes the two have remained associated, though in other cases the silvicultural system has been copied and combined with a different method of yield regulation or vice versa.

Conditions, not only of forests and climates, but also of staff, facilities for management, intensity of working, etc., vary so much that the most suitable method of management, including silvicultural treatment, assessment of the growing stock, fixation of the yield, and regulation of felling, can only be worked out from general principles and local knowledge on the spot.

In order to facilitate the organization of a forest or WORKING PLAN AREA, which is defined as the area covered by a single working plan, it is usual to start by forming COMPARTMENTS. A compartment is a territorial unit permanently defined for purposes of description and record. It is the smallest permanent unit of management, location of work, and record. As such, its boundaries must be carefully chosen and marked so that they will not be changed by accident, and so that no one will be tempted to change them by design. The boundaries should therefore be either natural features not liable to move, such as ridges, valley bottoms, rivers, where they can be relied on not to change course, or artificial lines such as permanent roads or rides (but not paths which are liable to deviate from their original course) or cleared traces, such as fire-lines. As the accessibility of compartments is a matter of great importance, and it should be possible to bring produce out of any compartment without taking it through another, it is often convenient to make the lower boundary of a compartment on sloping ground a road, or at least a suitable road trace.

It will be a great help to management if each compartment can be potentially a silvicultural unit, that is to say of uniform site[1] qualities as regards soil, exposure, aspect, etc., which will give the same response to the same silvicultural treatment all over. It is desirable, though of less importance, that each compartment should carry now one forest type only, suitable for descriptive inventory and uniform treatment. If it does not, temporary division into SUB-COMPARTMENTS, which are then the silvicultural units of management, can be carried out and the sub-division can be abandoned later when the crops have been brought into a condition when they do not require different treatments. This, of course, will only be possible if the site is homogeneous.

The size of compartments depends on the intensity of management which is possible for the forest in question. Naturally the smaller the compartment the easier it will be to include only one type of site in it. In a closely-worked production forest compartments can be quite small whatever silvicultural system is used. Even under the selection system, when each compartment must contain a complete range of size classes and is often the unit for calculation of periodic increment, it has been found in Switzerland that areas ranging from $12\frac{1}{2}$ to 25 acres are suitable. In their even-aged plantations the British Forestry Commission have adopted 25 acres as a convenient average size.

In large protection forests, where maintenance of good ground cover is the primary objective and working is likely to be light, larger compartments are usual. One of the most important considerations in such forests is the layout of roads by which fire-fighting gangs and materials can be moved, and compartments will often be primarily units of location and protection bounded by roads and fire-breaks. As will be seen later compartments are used like the pieces of a jigsaw puzzle to build up larger management units which are often only temporary. Small compartments give greater scope in build-

[1] See page 64 for discussion of sites.

34

ing up these units to the required size and for making adjustments between them. As the compartment is the smallest permanent unit of management it should not be too large to be covered by any treatment in one operation. Under a clear felling system it should be possible to clear the whole of a compartment at once without fear of exposing too large an area, or of not being able to replant it at one time.

On the other hand, the division of a forest into a large number of small compartments is expensive and can only be justified by the expectation of considerable production from them at reasonably short intervals. In a tropical forest from which only two or three trees per acre may be cut every 30 or 40 years, and where rides and cut lines become overgrown very quickly, permanent maintenance of large numbers of small compartments is not economic.

Also it must not be forgotten that the compartment is the unit of record. Recording requires the maintenance of several forms for each compartment, and a large number of compartments will mean a great deal of office work. Only considerable production and therefore revenue can justify the staff required for this.

In the tropics then, where most of the forests can only be worked extensively at present, and where a few trained foresters with inadequate clerical assistance have to deal with large areas, it is quite impossible to divide up great tracts of jungle into small permanent compartments and to maintain records for each. In Nigeria natural forest compartments of a square mile each are formed by lines cut at right-angles to each other. In Uganda all forest of a similar vegetation type contained in a ten-year working area is treated as one compartment, which may comprise as much as 5,000 acres. In both these territories, however, compartments in plantations are very much smaller because they are worked at the normal intensity, and when more complete working of the natural forests becomes possible it is likely that the original compartments will be split up. In America compartments are sometimes 40 acres each, but more usually 160 acres, and sometimes in remote forests may be as large as 5,000 acres.

When permanent division into compartments has been carried out any other division of the working plan area required either permanently or temporarily for management can be made by forming groups of compartments. If different systems of silviculture are going to be used in different parts of the area, different sets of working rules, called prescriptions, will have to be drawn up for the different parts. Such parts are known as WORKING CIRCLES and a W.C. is defined as 'The area (forming the whole or part of a working plan area) organized with a particular object and subject to one and the same silvicultural system and the same set of working prescriptions. In certain circumstances working circles may overlap.'

Then if it is considered undesirable for silvicultural reasons to have unduly large felling areas in any one place, or if it is wished to supply different contractors from different parts of the working circle, the circle may be divided into FELLING SERIES. A SERIES is defined as 'A management unit in which the object is to maintain or create a normal representation of age gradations', and a Felling Series as 'A series with a separate calculation of the yield, formed to control felling and regeneration'. When a Working Circle is not divided it is, of course, one Felling Series. (A Planting Series is similarly a series formed to control planting, and may be constituted for the afforestation of bare land in a working circle.)

Fig. 5 shows diagrammatically at (a) a working circle containing only one felling series of forty age-gradations in equal ANNUAL COUPES (annual felling areas) whose size is the area of the circle divided by the rotation, in this case forty years. The small profile diagram (a) shows the arrangement of the age-gradations, but only for the first ten years to save space. It should be noted that cutting should take place against the direction of the prevailing wind so as to produce the kind of profile shown with the smallest trees on the windward side when the series has been established. The younger age classes thus break the force of the wind for the older ones behind, and the young exposed age class develops strong root systems to resist wind when it comes to maturity. It will suffer some wind damage, but it will be protecting the crops which are felled behind it, and when these are exposed by the felling of the windward coupe a sloping gradation will be presented to the prevailing wind again.

The diagrams marked (b) in Fig. 5 depict in plan and pro-

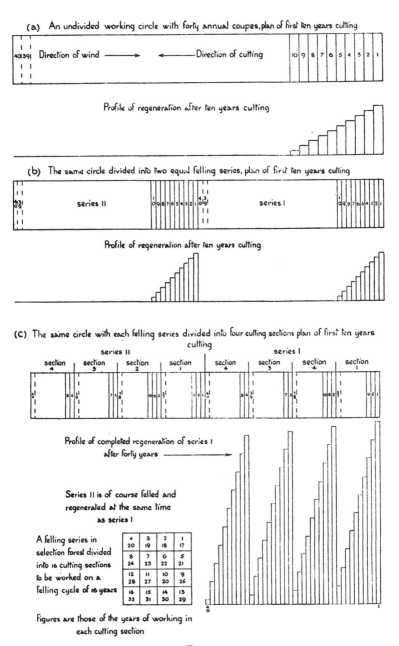

(a) An undivided working circle with forty annual coupes, plan of first ten years cutting

Direction of wind ⟶ ⟵ Direction of cutting

Profile of regeneration after ten years cutting

(b) The same circle divided into two equal felling series, plan of first ten years cutting

series II series I

Profile of regeneration after ten years cutting

(C) The same circle with each felling series divided into four cutting sections plan of first ten years cutting

series II series I

section 4   section 3   section 2   section 1   section 4   section 3   section 4   section 1

Profile of completed regeneration of series I after forty years ⟶

Series II is of course felled and regenerated at the same time as series I

A felling series in selection forest divided into 16 cutting sections to be worked on a felling cycle of 16 years

4	3	2	1
20	19	18	17
8	7	6	5
24	23	22	21
12	11	10	9
28	27	26	25
16	15	14	13
32	31	30	29

Figures are those of the years of working in each cutting section

FIG. 5

file a working circle divided into two equal felling series which are worked simultaneously for sustained annual yields. There is no need for felling series to be equal and a W.C. may be deliberately divided unequally. Sometimes it is desirable to avoid felling in adjoining sites consecutively for some particular silvicultural reason. This may be because of the fire danger presented by a large stretch of young coniferous woods, or because bark beetles breeding in the stumps of felled pines would attack young pines planted in the vicinity during the few years before the beetle population falls again. In such cases a felling series may be divided into CUTTING SECTIONS, which are defined as 'Sub-divisions of a felling series, formed with the object of regulating cuttings in some special manner'. The diagrams marked (c) in Fig. 5 depict the division of a felling series into four cutting sections with the annual coupe located in each in turn and returning to the same section only at four-year intervals.

These management units have been described with reference to the clear felling system, which is the simplest form of silviculture to understand. Under the uniform or shelterwood system age classes take the place of age gradations and periodic coupes take the place of annual coupes. Thus, in Fig. 5 (a) the first ten annual coupes or age gradations might be grouped together to form a periodic coupe (PERIODIC BLOCK) to be cut and regenerated gradually by seeding, secondary, and final fellings during a period of ten years. This would result in a periodic block of the 1–10 age class, and at the end of a forty-year rotation the whole working circle could similarly have been converted into four periodic blocks containing respectively the age classes 31–40, 21–30, 11–20, and 1–10. The number of periodic blocks in a working circle or felling series is found by dividing the rotation by the regeneration period, and the normal area of a periodic block is the area of the felling series divided by the number of blocks. The regeneration period varies, of course, according to the species to be regenerated, the climate, and the soil conditions of the site. It is purely a function of silviculture, and generally varies from 15 to 40 years.

In irregular forest with selection silviculture it is not possible or even desirable to work over the whole felling series (unit of yield control) every year. The area is divided into parts, also known as cutting sections, each of which is worked at an interval of years known as a FELLING CYCLE and defined as 'The time which elapses between successive felling operations on the same area'. Cycles may be of any number of years from about 6 to 50, depending on the intensity of working which is possible. If ten years is chosen a tenth of the felling series will be worked over each year, and, if the rotation is 100 years each part (cutting section) will be worked over ten times in the rotation, taking out in theory the trees 95 to 105 years old and thinning clumps of younger trees. In practice rotation means very little in a selection forest and the trees which are felled at each visit are those which have finished their period of maximum growth, are interfering with potentially more valuable trees, or require removal for other silvicultural reasons.

It must be realized that, though the organization of a forest has been explained by diagrams showing rectangular and equal areas in regular sequences, forests do not grow and regenerate evenly and regularly. Each stand needs individual treatment, and by appropriate organization can be given the necessary treatment at approximately the right time while still the overall annual output is kept at a fairly steady level.

Some form of area or volume control of silvicultural operations is necessary to prevent violent fluctuations of production and of demands for labour. While regular yields are often one of the objects of forest management there are cases in which considerable elasticity is desirable, for instance, in private estates from which death duties have to be met at unknown intervals, and in all forests to meet war demands and slumps. It may, therefore, be necessary to build up reserves of uncut increment or to decide where cutting of capital shall take place temporarily in emergencies.

Storing increment and so increasing the forest capital resolves itself into lengthening the rotation which usually means obtaining less increment on the capital, though the value of this increment may rise with the greater size of the trees cut. It will often be possible to reduce temporarily the rotation of selected stands, preferably those making the poorest use of their sites, and so increase immediate output. Similarly in times of poor demand and low prices reserves can be built up by delaying the felling of particularly healthy crops. The advantages of planned organization and recording are not confined to maintenance of sustained yield, but include guidance on how best to adjust the possible output to year to year demands.

PART II

PLANNING MANAGEMENT

CHAPTER

4

A WORKING PLAN is a written scheme of management aiming at the continuity of policy controlling the treatment of a forest. A forest in this definition means 'An area which for the most part is set aside for production of timber and other forest produce, or which is maintained in order to exercise climatic or protective influences on adjoining areas'.

Working plans may therefore have to be drawn up for very differing kinds of area, for bush growth on a water catchment, for bare land which is to be afforested, as well as for tracts which are to be managed for the production of any form of forest produce from gum to timber. The Americans speak of multiple-use forestry, by which they mean forest management for many different objectives combined in such a way that the achievement of none of them interferes too much with that of the others.

Essentially a working plan is the simplest possible statement of what is known about a working plan area, its configuration, soil, and climate, what is growing on it, and its possibilities, what has been done on it in the past, what should be done in the future, how it should be done, and what records should be kept.

In Great Britain and many other countries in which afforestation is being carried out one of the most important forms of planning is the planting plan. This does not differ in essentials from any other working plan. It should describe what is known about the area, its soil, climate, and vegetation, state what are the objects to be aimed at by forming plantations, what is to be planted where and when, how it will be protected and tended, along what routes thinnings and later other produce will be taken out, and what markets it is intended to supply. If no plan is made species may be put out on unsuitable soils, age-class distribution may be abnormal, cleaning and thinning

43

work may accumulate in excess of the capacity of the labour supply, crops may grow up in areas from which they cannot be extracted economically, or may produce material which is not readily marketable.

There are different opinions about what should be the unit of a working plan area, an individual forest or group of forests which can be worked together, or the administrative charge of a forest officer, such as a conservancy or division. In Europe the individual forest is generally the unit, but in India there was a tendency to combine the forests of a division into a plan with numerous working circles. However, in Great Britain and all the colonies there are so many plans to be drawn up that they will have to be made first for any areas which can be tackled, usually individual forests or groups of forests, and the question of possible combination can be left for consideration in later revisions.

There are two schools of thought about who should draw up plans, a special branch of working plan officers or the normal territorial staff. The argument for the local officer is that he should have a much more detailed knowledge both of his forests and the needs of his districts than anyone else, and should be able to plan the most practical organization for the general good. This is certainly true in Europe, where foresters stay for long periods in one area, but may not apply so much in the tropics. However, the local man will have to carry out the plan, and there is something in the argument that, in order to ensure that plans shall be clear and workable, no one should write a plan which he will not have to operate. It is argued, on the other hand, that local officers are so busy with routine work and administration that they have not the time to devote to planning, whereas specially experienced working plan officers can give uninterrupted attention to the work and so cover large areas in the minimum time.

Probably a combination of both methods will give the best results. Either the local officer can be given the temporary services of special staff to collect data, prepare maps, etc., or a special working plan party can collaborate closely with him. In any case, there will have to be very full consultation and discussion about objects of management, silvicultural methods, etc., before a plan is made, and in a State forest department each plan should represent the considered opinion of the local officers, their superiors, and any specialists available, the final decisions being made by the head of the department. The head of a large department often delegates the supervision of planning to a conservator in charge of a working plans section or branch. This officer arranges for the preparation of new plans and the revision of existing ones, either directly by his own staff, or by giving assistance to territorial officers. He reviews all the plans made, checks their annual progress from the control forms, and keeps the head of the department advised of the general position.

In a private forest the owner must decide the objects of management after consultation with his forester, estate agent, keepers, etc., and possibly with the advice of a consultant forester. The forester or the consultant will then draw up a plan to give effect to the owner's wishes as effectively as possible.

Whoever draws up a plan, it must be submitted to the owner or, in the case of a State forest, a high official of the State, who after consideration and possibly having some amendments made, will sanction it formally, and prohibit deviations from its prescriptions without his authority in writing. In the case of French State Forests a Decree is issued over the signature of the President of the Republic and that of the Finance Minister in respect of every state forest for which a plan is drawn up. The decree states the period for which the plan is sanctioned. In order to ensure reasonable elasticity minor deviations of timing and technique are allowed to be made by the executive officer, and rather wider powers are granted in the plan to officers in charge of regions, while revision must be possible at any moment, if the need for it becomes apparent before the sanctioned period is ended. The formality of the procedure must be sufficient to deter attempts at alteration without weighty reason. In fact, red tape is evoked to fulfil its proper function of allowing time for consideration of new and enthusiastic proposals.

If something like this is not done there will be constant changes of intention, species, methods, etc., because every officer who takes over a forest will want to try something different and there will be nothing to stop him. The present state of the Forest of Dean is a good example of this, because though several plans have been drawn up for it, none has ever been sanctioned, and almost every new Deputy Surveyor has put his own ideas into practice.

Private forest owners in Great Britain have considered on the whole that working plans are too elaborate and rigid and that their forests can be managed best by drawing up annual plans of operations which can be followed easily by semi-trained staff. They have not realized that plans could be drawn up which would allow for large intermittent fellings to meet death duties and for adjusting annual output to the state of the market. If these things form part of an owner's objects of management foresters must find ways of giving effect to them while still safeguarding the soil and organizing good silvicultural treatment. There must be some agreed general scheme of working towards an aim, or even alternative aims, and it is just as easy to draw up annual schemes of operations to fit into the general scheme for several years ahead, with revision on alternative lines in emergencies, as it is to work out every year afresh what should be done in the next.

During the recent war, when demands for increased output of timber reached India, foresters there were able to refer to their working plans and control records, to see first where fellings were in arrears, and then where future exploitation could be accelerated with the least damage to the forests. The result was that, though heavy cutting took place, its effects could be redressed by post-war revisions of the plans in a period of something like ten years. In the colonies, on the other hand, foresters had practically no plans to which to refer. They had to search hastily for any stands of timber which could be exploited, without time to investigate all possible alternatives, or to estimate the effects of the felling on the future of the forests. Much permanent harm was done in consequence.

In many colonies it has been argued that working plans cannot be made before forest reservation has been completed, forests allocated to their main functions, demarcated and surveyed, enumerated, and classified into types. These things have not been finished, but timber has had to be cut for local use, or for export to obtain revenue, and quite a lot of planting has been done. Neither of these operations should be undertaken without plans, or the result is chaos. Lack of good maps, of knowledge of forest types and their silviculture, ignorance of the volumes of growing stock and rates of increment all mean that only very tentative plans can be made, but these can prescribe continuity of records, gradual survey and demarcation, and the collection of essential information. There must be some basis on which cutting is allowed to start in a forest, if it is only that not more than a certain number of trees a year are to be cut. If this is put on paper and provision made for keeping a few simple records a scheme of management can be evolved in time, or at the very worst those who follow after will not be left in complete ignorance of the early history of the forest.

A plan to make a working plan is better than no plan at all, and once the ice is broken progress will be made.

The WORKING PLAN PERIOD is defined as 'The time for which detailed prescriptions are laid down in a working plan'. There must be some general basis for long-term development in a plan. For instance, if a forest is to be worked provisional decisions will have to be made about lengths of rotation or the size of produce it is intended to grow, whether sustained or intermittent yields will be the objective, etc. On the other hand the idea that the whole future of a forest could be mapped out and ordered for very long periods has proved to be fallacious, because nature cannot be regulated to machine-like precision. Good seed-years may come at irregular intervals, droughts may interfere with regeneration, storms may blow down trees before they were scheduled to be cut, and regeneration may occur in areas where it was not intended. Besides all this, markets fluctuate and the sizes or kinds of produce in demand may vary.

It has therefore been found unwise to plan for long periods of a hundred and more years as some early foresters did. On the other hand, revising a plan takes time and labour which should not be expended too frequently. Sometimes there may be some inherent factor of the management prescribed which will make some particular period an appropriate one, for instance, a regeneration period of twelve years suggests planning and revision at twelve-year intervals, but a twenty-year regeneration period might well be split into two ten-year working plan periods. Ten years has often been found convenient

47

for detailed planning, five years is probably the absolute minimum. Collection of material for revision should start a year before the expiration of a plan, and it is undesirable to have to do this in the fourth year from the start. A general review of progress at regular intervals provides valuable information about rates of growth, progress of regeneration, development of markets, etc., even if no change in the general scheme of management is necessary. Where tentative provisional plans have been made originally they can be built up and improved on the basis of information recorded during each period.

It is very useful to have a standard list of headings for the compilation of a working plan, not so that something shall be dragged in under every heading, but so that nothing shall be omitted by oversight instead of by intention. It is also helpful that the facts upon which the plan is based and the proposals for the future should be presented in a logical sequence. If this sequence is always the same an officer taking over a new area will be able to understand the plans which have been made for it more easily than if they were in unfamiliar forms.

The late Mr. Ray Bourne drew up such a list and published it in the *Empire Forestry Journal*, Vol. 13, of 1934. This received very general approval and was adopted almost as it stood by Mr. M. R. K. Jerram for his textbook *Forest Management*, published in 1935. It is certainly not the only possible arrangement, nor the only one which would be suitable, but it is logical and comprehensive and is therefore given below with a few minor amendments only. The arrangement can be used for elaborate and detailed plans, such as have been drawn up in India, but equally it can form the basis for development plans for areas about which there is little information, and for simple plantation plans of a few pages of typescript.

This is how Bourne explained his choice of matter and sequence. 'Experience in the examination and compilation of Working Plan Reports has conclusively demonstrated the desirability of presenting the matter in such a sequence that forward reference and unnecessary repetition are eliminated. The sequence followed in the attached statement of suggested Working Plan headings has been found very suitable and the reasons for its adoption may be explained.

48

The objects of dividing the Report into two parts—Part I: Summary of facts on which proposals are based, and Part II: Prescriptions for future management—and providing for an Introduction and Appendices, are so obvious that further comment would be superfluous. The significance of the suggested division and arrangement of Part I, however, will not be appreciated as readily. Following upon a statement of location and ownership in Chapter I, the local conditions are described in Chapter II. Since the configuration may materially influence the climate, particularly the local climate, the topography is first considered. Then, since climate has a great effect on the weathering of rock and the development or erosion of soils, this subject is discussed prior to the geology. It may be noted here that, in the description of the topography, reference should not be made to the geology, a special section being devoted at the end of the chapter to a statement of the relationships existing between topography, climate, geology, and soil. It may again be noted that in the description of the soils, no reference should be made to the influence of the vegetation on soil processes, a special chapter, No. IV, being provided for the purpose. In Chapter II, simple descriptions of the several soils will suffice. A discussion of the effects of the vegetation on the soils would be premature, since the effects of human intervention in the past on the natural vegetation of the area must first be considered and, for this reason, Chapter III is devoted to an historical statement. In the first section of Chapter III, historical descriptions of the vegetation should be given and brought up to date, while in the second section the injuries to which the crops have been subject in the past should be described. It is also logical and reasonable in this chapter to refer to the works of improvement undertaken in the past and to past yields. In Chapter IV the ecology of the forest can then be reviewed in consideration of the locality and biotic factors previously stated and described. The future treatment of the different crops comprising the forest may also be discussed from the scientific aspect. In Chapter V any facts necessitating the modification of scientific principles in practice, on economic grounds, should then be put forward. Finally, in Chapter VI, the data collected with a view to the regulation of the yield and the methods employed should be reviewed. In this manner all the facts on which the future prescriptions should be based can be presented in logical sequence and the reasons for such prescriptions elucidated. In Part II there should be no need to adduce further reasons for the adoption of a particular policy or treatment. The prescriptions laid down should obviously follow from the facts set out in Part I.

In Part II, which should be kept as concise as possible, the general plan for the entire forest is to be outlined in Chapter I. On the assumption that the areas set aside for the fulfilment of different objects will constitute separate Working Circles, a chapter is then devoted to the detailed prescriptions for each Working Circle. If a Working Circle is divided into two or more Series, the special plans in Sections 5 and 6, which should, if possible, be drawn up in tabular form with suitable control columns, will have to be repeated for each Series. With regard to the records to be made during the execution of the plan, it may be noted here that provision is made in the Compartment Register for the detailed record of work done in each Series and Working Circle by Compartments. Such record is maintained by unit areas. Annually the records should be summarized by subject and the summaries posted in the control columns of the special plan for each Series or Working Circle. Estimates and actuals of receipts and costs should also be posted in the special plans to furnish the basis of annual plans of operations and budget statements. Finally, the last two chapters of Part II are devoted to miscellaneous prescriptions applicable to the forest irrespective of Working Circles, and to a financial statement of the present and probable future positions.

The number and nature of the appendices and maps required will vary according to circumstances. In all cases a Compartment Register and a Stock Map are essential. If possible the statistics of growing stock, age-classes, quality-classes, size-classes, etc., should be included in the Compartment Register, which in turn should be in a form convenient to carry in the field. Finally, experience has shown that the actual forms to be used are most conveniently drawn out and especially adapted in each case, or at least to each method of management.'

ARRANGEMENT OF A WORKING PLAN

The headings given here are neither exhaustive nor suitable for all cases: any not required should be omftted. A Working Plan Report should be brief and to the point. Detailed figures, etc., should be relegated to Appendices.

TITLE

Name and period of plan, author's name, date.

REFERENCE TO RECONNAISSANCE REPORT AND CORRESPONDENCE

Quote general recommendations as to Policy and treatment and include correspondence relating to the preparation and sanction of the Plan.

TABLE OF CONTENTS

Reference to pages and paragraphs, both of which *Should Always be Numbered Consecutively Throughout.*

SUMMARY OF PRINCIPAL PRESCRIPTIONS

Division of the area into Working Circles and Series; Species; Rotations or Exploitable Size; Silvicultural Systems; Methods of Yield Regulation; Periodic Yields; Period of the Working Plan.

PART I

SUMMARY OF FACTS ON WHICH PROPOSALS ARE BASED

CHAPTER I. LOCATION AND OWNERSHIP

Section: 1. *Name, situation* Under *situation,* enter name of county, civil district, forest division, or other administrative unit.

2. *Distribution and area* Under *distribution,* state whether in one block or more and whether compact or scattered. *Area* thus:

Potentially productive		626 acres
Permanently unproductive:		
Roads and rides . . .	22 acres	
Grass-lands . . .	56 ,,	
Waterways . . .	5 ,,	
	—	83 ,,
	Total	709 acres

3. *Boundaries:* N.E.S.W.: nature of boundaries and general state of boundary marks.

4. *Ownership and legal position:* under *legal position* include a brief summary of rights and privileges. Under *ownership,* if state-owned merely refer to 'The State Forest of ....'.

CHAPTER II. LOCAL CONDITIONS

1. *Configuration:* flat, undulating, hilly, mountainous; main drainage basins and watersheds; main aspects of hill slopes; existence of precipitous unworkable ground.

2. *Climate* Rainfall and its distribution, snowfall; atmospheric humidity; temperature; occurrence of frost and drought; prevailing winds.
3. *Rock and Soil* Geological formations. *Soil:* composition, depth, porosity, degree of moisture, structure; descriptions of sample profiles may be given.
4. *Sites* A summary of sections 1–3, listing the principal sites, and if convenient grouping them by regions.

CHAPTER III. HISTORY

1. *General history of forest* A review of events which have influenced the species and crop forms in the past; past systems of silviculture and management; dates of previous Working Plans and the consequent divisions of the area. Critical review of results of past management.
2. *Injuries to which the crops have been liable* Animal agency—insects, rabbits, squirrels, deer, cattle, etc. Plants—fungi or other parasites, weeds, climbers. Climatic—frost, snow, hail, wind, drought. Other forms of injury—fire, theft or damage by man, etc.
3. *Works of improvement undertaken* Buildings, roads, draining, etc.
4. *Past yields* A summary only, reference being made to the documents in which the details are recorded. Volume yields to be separated from figures of revenue and expenditure.

CHAPTER IV. ECOLOGICAL CONSIDERATIONS

Forest Types Sample descriptions to be given of typical crops and their future discussed. A key to be prepared, setting out the position of each type in a general classification.

CHAPTER V. ECONOMIC CONSIDERATIONS

1. *Requirements of surrounding population.*
2. *Markets and marketable products: price movements.*
3. *Lines of export, methods of exploitation, and their cost.*
4. *Staff and labour supply:* availability, efficiency, and cost.

CHAPTER VI. STATISTICS OF GROWING STOCK AND INCREMENT

Procedures followed in the collection of data should be carefully described, and tabular summaries of the areas and volumes of the stock by species and age classes with increment rates should be included when possible. Details should be left for an Appendix.

PART II
(To be kept as concise as possible)

PRESCRIPTIONS FOR FUTURE MANAGEMENT
CHAPTER I. GENERAL PLAN

1. *Objects of management* The final definition of the policies to be pursued.
2. *Outline of general organization to be adopted.*
3. *Division of the area* (a) Names and areas of working circles and of systems of management prescribed. (b) Principles on which compartments constituted or revised.
4. *Period of working plan.*

(One chapter to each working circle)

Note If there is only one working circle a separate chapter is not required, the sections enumerated below being placed in Chapter I.

1. *Constitution of working circle* In one compact block or in scattered blocks; situation; names of local blocks or compartments included, and other particulars.
2. *Sub-divisions of working circle* Series, cutting sections (if any), compartments, etc.
3. *Species. Silvicultural system and rotation or exploitable size* Prescriptions as to technique of thinnings and regeneration fellings, etc.
4. *Yield regulation*
   (*a*) Division of rotation into periods or felling cycles, etc.
   (*b*) Estimate of growing stock; distribution of age classes.
   (*c*) Rate at which regeneration should proceed.
   (*d*) Calculations of annual possibility.
5. *Plan of exploitation*
   (*a*) Final cuttings
   (*b*) Intermediate cuttings } Tabular statement of cuttings, estimated value, and cost (or vol.).
   (*c*) Minor produce
   (*d*) Felling rules
   N.B. Control columns to be provided.
6. *Plan of cultural operations* Areas to be planted up or otherwise regenerated each year or period—cost. Weeding, cleaning, fencing, draining, closure to grazing, etc.—cost.
   N.B. Control columns to be provided.

CHAPTER . . . MISCELLANEOUS PLANS

Prescriptions applicable to whole area, if not already entered under separate working circles. Control columns to be provided where necessary.

Examples of possible sections:

1. *Development of forest industries* Cost.
2. *Roads, waterways, and other export works* Cost.
3. *Fire protection plan* Cost.
4. *Drainage plan* Cost.
5. *Buildings* Cost.
6. *Other works* Additional survey, mapping, and re-enumeration. Demarcation or repairs to boundaries, etc.—cost.
7. *Research* Programme of experiments, etc.—cost.
8. *Establishment and labour* Proposals for revision of strength of establishment; questions of labour supply—cost.
9. *Control of working plan* Methods of recording work done and of analysing results for purposes of control.

CHAPTER . . . FINANCIAL

1. *Estimate of capital value of forest* Not always possible.
2. *Financial forecast* Estimate of annual receipts and expenditure under

different heads; net surplus for whole area and per acre. Interest on capital (if possible).

3. Cost of preparation of the plan.

APPENDICES

Should include, probably among others:

  I. *Description of Compartments.*
 II. *Results of enumerations of growing stock.*
III. *Allotment of areas to age-classes or periods.*
 IV. *Statistics of Volume, rate of growth, etc.*
  V. *Detailed calculations of yield.*
 VI. *Special control forms.*

MAPS

  I. Index Map
 II. Geological and Soil Map
III. Stock Map
 IV. Fire Control Map, etc.

If reproduction is proposed, these maps should be combined for reasons of economy.

Management divisions can be shown on the Index Map, or a separate Management Map can be prepared for use as a record of progress of exploitation and regeneration.

The individual aspects of management will now be dealt with in detail in the order in which they appear in this list of possible headings for a working plan.

The first item reads '*Reference to Reconnaissance Report and Correspondence*' and here right at the beginning is something which does not appear in every plan. A reconnaissance report is a preliminary report submitted to an owner, government or other, based on an examination of a forest area made in order to gain a general knowledge of all the facts likely to be helpful in determining what use can be made of the area, and what sort of management should be applied to it. The report may well contain alterative proposals for the general policy to be pursued and the probable advantages to be obtained from each. The owner makes his decision and issues instructions for the preparation of a plan to attain certain objects of management. This procedure was customary in India, where the report was called a Preliminary Working Plan Report and, after discussion, often formed the basis of instructions given to a working plan officer. In order to form a reliable guide such a report should really cover all the points which are dealt with under Part I of a plan, and it has often been found that some of the assumptions made as a result of a rapid reconnaissance are not fully borne out by the detailed collection of data carried out later. In such a case the instructions issued would possibly have to be amended during the preparation of the plan. The real value of a reconnaissance lies in determining the manner in which collection of data shall be started, as for instance the intensity of sampling which is at once desirable and practicable.

In many cases no preliminary report is needed. Most forests in which any work has been done have been managed with some sort of objective, possibly rather vague, and on lines which have probably varied, but which will have provided a certain amount of information about their potentiality and silviculture. For such areas, and even for some forests which are to be opened for working for the first time, it may well be that the preliminaries will take the form of discussions between the executive officer in charge, his seniors, such as a divisional forest officer and the head of his department, and any specialists available, such as a silviculturist and a forest engineer, during a joint inspection of the forest. The officer who is to make the plan, if not one of those already mentioned would, of course, be one of the party, and agreement would be reached about the general lines on which the plan is to be based, and on the manner in which the work is to be tackled. The senior officer would then make a record of the decisions and issue it as an instruction to the planning officer. If in the course of his work this officer found that any of the assumptions which had been made were not confirmed by his detailed investigations, he would report the facts and after further discussion his instructions would be varied if necessary.

It is, therefore, clear that what will be included as an introduction to a plan will depend on circumstances. Very often it will consist of a copy of the letter of instructions issued, of any later amendments, and of the final sanction of the plan. French plans always open with a copy of the Decree signed by the President, to which reference has already been made. In Indian plans reference is frequently found to any alterations which were made to the original plan as submitted by the working plan officer, on the instructions of the senior officer, who checked it before recommending sanction. Often they also contain copies of minutes by the administrative officers of the district and province, certifying that all rights of the local population have been adequately provided for, and that there is no administrative objection to any of the clauses of the plan.

Table of Contents The double indexing by page and paragraph is useful. Consecutive numbering of paragraphs throughout the plan is really important in a document of which the paragraphs are likely to be the subject of correspondence. It is very much easier to refer to and to find paragraph 200 than Part II, Chapter 4, paragraph 5.

Summary of Principal Prescriptions This cannot be drawn up until the plan is finished, and so is sometimes forgotten or put at the end. It is, however, very important, and the best place for it is the beginning, where it gives the reader the gist of the plan, and so helps him to follow it as it is unfolded. Nothing more will be said about it here, but when all the matters which have to be summarized have been dealt with, the various forms which the summary can take will be discussed.

CHAPTER

5

Name and Situation Besides the administrative location (county, district, or other local unit) the latitude and longitude limits and the geographic position, such as 'in the valley of the Blank River', should be given. The forest district, division, or other administrative charge in which it lies should also be stated.

Distribution and Area Block, in the sense used by Bourne, means 'a natural main division of a forest, generally bounded by natural features and bearing a local proper name', and should not be confused with a periodic block, which is a periodic coupe, as opposed to an annual coupe. In a working plan area blocks may be separate forests or portions of one forest divided by rivers, etc. These blocks should be listed in a table with their potentially productive and permanently unproductive areas, unless there is a very large number of them, in which case the total figures should be given in this section and reference made to an appendix in which the full details will be found.

This raises at an early stage the important question of the distribution of matter between the text of a plan and its appendices. There can be no fixed rule about such distribution, but a useful guide can be obtained from remembering that it should be possible for any one, including particularly a layman such as an administrative officer, to read and follow a plan without turning to the appendices and without finding the text cluttered up with a mass of detail and figures. A forester who wants to study the detail should be able to refer to various appendices in which he will find it, and should be told at the appropriate places in the text which appendix he should turn to.

Bourne, in his layout, lists roads and rides among the permanently unproductive areas, but unless very intensive forestry is being practised it is quite legitimate to consider rides and forest roads, as opposed to main or public roads, as part of the potentially productive area. The crowns of trees will meet over many rides and presumably the roots beneath them. The productive area of each compartment is often measured from the centre of the rides which bound it, and the sum of the areas so measured deducted from the total working plan area gives the total unproductive area. Except in very intensive work areas should be measured to the nearest acre only. However, in the British Forestry Commission Dedication Scheme areas to the nearest tenth of an acre are asked for. In this matter, as in many others to be dealt with, it must be remembered that some forestry is necessarily akin to ranching and some to market gardening.

Boundaries The legal outer boundaries of the area or its individual, isolated portions should be recited here, unless this would take up a lot of space. If there are a large number of boundaries it will be better to insert some general remarks, such as 'The boundaries are clearly defined on the ground by hedges, stone walls, or fences, all of which are in good repair and are listed with their directions, lengths, and state in Appendix I'.

Often there is a legal description of the boundaries, such as a title deed or notice gazetting a forest reserve to which reference can be made and of which a copy can be appended, but notes on their physical state, visibility, and effectiveness in preventing trespass at the time of making the plan should be added if of importance for the protection of the forest.

It is usual to describe boundaries starting from some easily identifiable point on the ground, if possible a survey beacon or one located by its distance and bearing from such a beacon, in the north-west corner. The description, preferably in tabular form, proceeds in a clockwise direction by legs with distances and True North bearings (with a note of the date and method of correction from Magnetic North) back to the starting point.

When a forest is demarcated for reservation the choice of boundaries depends very much on local circumstances. For instance, if it is a case of securing a provisional reservation in country sparsely inhabited by a nomadic tribe, the object will be to choose boundaries which will not need maintenance, but can be identified and modified later when the country develops. Permanent rivers, crests of ridges, or even compass bearings from one hill-top to another may be used, but not tracks and paths which are liable to be altered. Later it is probable that bays of agricultural land inside these long lines will be excluded, and other adjustments made involving a number of shorter boundaries, which will have to be clearly visible on the ground to the local population. Permanent rivers, roads, and railways will be used wherever possible to keep down expense, but ridges, etc., will have to be defined by cut lines marked by permanent, intervisible pillars or cairns of stone or concrete. At all points at which there is a change of direction the position of the pillar should be fixed accurately on the map, and short trenches should be dug from it in the directions the boundaries lie. These pillars and, on long straight lines, some intermediate pillars, may well be given permanent serial numbers which are also shown on the map. The lines will require cleaning at intervals and some of them may be cleared to serve as fire-lines. Another way of making them obvious is to plant or sow some fast-growing, hardy tree species, generally exotic to the district, at intervals. Later still may come the stage of hedges, fences, and walls common in heavily populated countries.

Many forest departments have their own local standing orders about the selection and description of forest boundaries. *Ownership and Legal Position* Ownership should be stated clearly with reference to title deeds, or number and date of the legal notice constituting the area as a forest reserve. Any rights and privileges to which the area, or any part of it, is subject should be mentioned. If these are at all complicated and numerous, as they may be after a forest settlement has been carried out in a populous area, they should be summarized in this section and full details should be set out in an appendix. These will include such things as lists of all those entitled to free firewood, with the amounts, the manner, and the places in which they may collect it. Rights are very important to a planner because they must be satisfied before any other disposal of the yield is considered, unless it is otherwise

59

expressly provided in the settlement of rights. Thus, the existence of rights may necessitate the formation of a separate working circle or cutting series, or even preclude some particular form of silviculture or conversion to a different form or type of forest in which the rights could not be enjoyed.

Fortunately no right can be exercised in law in such a way as to destroy its own existence. Thus, because if more than the sustained annual yield of a forest was removed persistently, the forest would be destroyed, a right to unspecified amounts of forest produce can be restricted legally to the sustained yield of the forest over which the right is held. A working plan, by establishing what this sustained yield can reasonably be expected to be, can restrict a right and save a forest from destruction.

CHAPTER

6

LOCAL CONDITIONS

The information for this chapter of the plan is collected from careful study of the area, from the local knowledge of those living in or near it, and from records, published and unpublished. Methods of study depend on the size and type of tract which is being dealt with, on the maps obtainable, and on the time and funds available for the study.

In the case of a large area marked as 'wooded' on a map which shows no interior detail, except possibly some formlines which have been sketched in, a preliminary reconnaissance is needed to investigate the topography and to decide in what detail it should be mapped. The usual procedure after this is to lay out a convenient base-line parallel to the main ridges and from it to run a series of sampling strips at right-angles across the contours at fixed intervals which, with the width selected for these sample strips, will give the intensity of sampling of the growing stock for which staff and time are available, or which is deemed to be adequate. Thus, if the strips are one chain wide, which is often convenient in thick forest, and their centres are ten chains apart, the area sampled will be ten per cent of the whole. Such a system of systematic sampling of the growing stock by strips is frequently adopted in unknown forests in order that mapping may be combined with enumeration of the stock. The strips are used as survey lines, on which are fixed the points at which rivers, changes of slope, changes of vegetation type, etc., occur. By digging soil pits at fixed distances or at selected points along these lines the soil, and changes in soil types can be studied also. With a certain amount of inspection between the lines the topography, forest type, and soil type distribution can be mapped, but if the intensity of sampling is low, for instance if the lines are a hundred chains (one and a quarter miles) apart, as they would be for a one per cent enumeration with strips one chain wide, the detail between the lines would be very sketchy. In such a case the working plan should prescribe for more complete mapping of the forest during the period of the plan, but Part I is not the place for any prescriptions. Those for the collection of further information of any sort should appear in the Miscellaneous Plans chapter of Part II.

It will be appreciated that when this method of systematic strip sampling is adopted no estimation of the probable error of enumeration of the growing stock is possible, but it would be difficult to combine random sampling with mapping without a great deal more line cutting per unit of area sampled than is necessary for strip sampling. In tropical forest it is the cutting of lines which is the expensive item, and it is the usual practice to enumerate the whole of every line cut, whereas in temperate forests random plots or lines selected at random can be located with ease.

In properly mapped, intensively worked forests already divided into compartments, but not yet under plans, such as are found in the British Isles, existing roads and rides facilitate preliminary inspection, during which the topography can be studied and decisions made about what investigation of the soils will be needed. Possibly a string or two of soil pits across the main contours supplemented by inspection with an augur between will permit the boundaries of the soil types to be mapped.

The extent to which detailed description will be carried depends on the nature of the tract, the time and staff available, and the intensity of working which will be possible. At one extreme is the case of a protection forest covering several hundred square miles of rugged mountain catchment, in which the only exploitation will be that of some poles and firewood from near the lower boundaries by right holders. Revenue will be nil and only sufficient information is required to enable prescriptions to be made for the prevention of deterioration of the soil cover and the water-regulating powers of the vegetation. At the other extreme is a small forest property in a highly developed area, where every bit of produce that can be grown can be sold at varying degrees of profit. There will be markets for saw timber, telegraph poles, pit-props, fence posts, pulp wood, turnery poles, bean sticks, and firewood, and also for Christmas trees, evergeen foliage, and moss. In such a case every yard of soil is valuable and the fullest possible information is needed in order to be able to make the most profitable use of it.

Configuration The general topography should be described, bearing in mind that altitude, slope, drainage, and aspect will each have an important bearing on management, both in respect of silviculture and extraction of produce.

Climate. It is rarely that meteorological data are available from stations actually inside the working plan area, and the records of stations nearby, but outside actual forest, may not be strictly true for the forest. It should not be forgotten that extremes of climate are often of more importance to the forester than the means. It may well be that the average rainfall for the last fifty years would be very suitable for a certain species, but that forty years ago and again twenty years ago there were drought years which that species could not possibly have survived. Again, it may be the weather at some particular time of the year, such as the number of frosts in May in the south of England, that can be a determining factor in the choice of silvicultural methods. Actual observation in and around the area will enable areas subject to wind damage, snow break, frost damage, etc., to be located and related to topography.

Geology and Soils The geological formations determine the parent material of the soils, subject to the past geological history, but it must not be forgotten that the soils now covering the area may have been transported from elsewhere. Differentiation of soils into types should be carried out on practical, not pedological, lines. What a forester wants to know is the relative suitability of the different soils for various forest crops and treatments. Physical conditions, depth, permeability, drainage, texture, colour, etc., are generally of more importance to him than chemical composition. Under undisturbed conditions soil types are reflected in the vegetation they carry, but where there is reason to suspect interference this cannot be relied on. Use should be made of any geological and soil maps available but these will probably be on very broad lines and will need supplementing by soil-pits and the use of an

63

earth augur in the manner already suggested. Information can sometimes be obtained from prospectors and local farmers. Some native peoples have evolved their own classifications of tropical soils according to the food crops that can be grown on them, and these should not be ignored.

Sites   The idea of the section Bourne devoted to sites is the correlation of the preceding factors of configuration, climate, and soil into locality sites which, in the absence of any biotic interference, would produce sufficiently different types or development of vegetation to be taken practical account of by a forester. At the moment biotic factors, such as interference by man, animals, or fire are not being considered. These may have caused different parts of what is essentially the same site to carry different types of forest vegetation, for instance, part of a site on which the natural climax is an oak-beech mixture may have been felled and planted with pine. Another part of the same site may have been felled but not planted and have been invaded by birch and other hardwoods. Thus, these three areas would appear very different, but given the same silvicultural treatment the whole site could carry one crop type which could be managed in the same way. The sites which it is desirable to differentiate are management sites, and the object of dividing the working plan area into sites is the practical one of being able to say that all parts of each site could be managed in some particular way to obtain a desired result.

In untouched forest, site and forest type should coincide, but even in remote tracts it is very dangerous to assume that they do. Part may have been farmed in the past and may now be carrying a different seral stage in the succession back to the climax type. Because of persistent fire or grazing another part may be unable to progress beyond a pre-climax. Nevertheless, the forester wants to know what parts of his area will, after the introduction of proper protection and tending, be able to produce healthy, good quality crops of any one type of produce. This he can decide by considering the combinations of topography, climate, and soil. For instance, there may be on the area several ridges, exposed at certain seasons of the year to hot, dry winds, and covered with rather a shallow soil. These could probably be assigned to one site which would have to be managed for the production of non-exacting, wind-firm species under a silvicultural system that never left the soil unshaded. Again, the sheltered valleys with deep, moist soil might be considered as one site but, if parts of them were liable to frosts and others were not, there might be two sites in them, because only frost-hardy trees could be grown on one.

By region, Bourne meant an association of sites which might be grouped according to topographical or geological regions, e.g. a Flood Plain Region and a Plateau Region, or a Limestone Region and an Old Red Sandstone Region. Unless a very large area is being dealt with there is no need to group the sites into regions. Except for very intensive forestry the object should be to differentiate as few, broad, simple sites as possible, each of which will be of practical significance in the management of the area.

HISTORY

General When a working plan is made for an area for the first time it is very important to discover and record all that can be found out about its past history. When plans are revised Part I normally records only the additional history since the date of the previous plan, and any revising officer is entitled to assume that all possible researches were made and all available sources tapped by the original planner. The history is important, because it may explain many of the causes for the present state of the soil and of the crops. Books and local records should be sought out and oral tradition should not be neglected. Once again, a good deal can be gained from careful observation. High forest of seedling origin can generally be distinguished from that of coppice origin, and if a crop has been planted the original lines can often be made out, in the case of oak even after a hundred and fifty years.

If there has been any management in the past the results and the lessons to be learnt from them should be discussed. In the event of a plan being under revision a critical review of the past period will be the most important part of this section. *Injuries to which the Crop has been Liable* Most of the possible forms of damage are listed by Bourne. These may be of great practical importance, e.g. rabbits may make it impossible to obtain natural regeneration without fencing at prohibitive cost. The actual animals which cause damage are, of course, different in various countries. In an African working plan it is stated that 'Elephants are a major pest in the regeneration areas . . . in herds of up to 400 strong'. Ivy and honeysuckle in this country and lianes in the tropics are examples of plant pests. The very luxuriance of weed growth may be a controlling factor in silvicultural technique. Only careful observation and understanding of what is observed will enable this section to be completed comprehensively.

Works of Improvement These should be obvious. Roads, tramways, bridges, houses, stores, loading ramps, water tanks, or dams for fire-fighting, and drainage works are among the improvements likely to be found.

Past Yields Records may or may not be available. Sometimes they may be difficult to find or to disentangle from those of other areas. It is easy to write 'no records available', but more useful to give rough estimates of the sort of quantities of produce of various kinds which has been taken from the area either for sale or use by right holders.

CHAPTER

7

ECOLOGICAL CONSIDERATIONS

Forest or Crop Types in the management sense are practical ecological types, which must be visibly different from each other either in composition or form, and must either require different silvicultural treatment, or yield different products. It follows that they can be mapped, though their boundaries will often be transition belts of which the middle will be indicated on the maps.

In untouched forest these types should be coincident with the locality sites discussed in Chapter II, Section 4; for instance, Swamp Forest caused by impeded drainage. More often they will not; for instance, Secondary Forest (regrowth of coppice and pioneer species) and Savanna Woodland (scattered fire-hardy species over grass) may be found on different parts of the same site after shifting cultivation of high forest, simply because one part has escaped fires and the other has been burnt recurrently. The real expression of the site in each case is High Forest. The same forest type may also be found on two different sites, as Pine Plantation partly on an Oak / Beech Forest site and partly on a Heath site. In this case, however, owing to the difference in quality of the two sites, the crop type might have to be split into two quality (productivity) classes which would need different thinning regimes.

At some stage in the reconnaissance or preliminary inspection a decision must be made about what types are going to be differentiated. No two forest stands will look exactly the same and, as already mentioned, any types selected are likely to have transition stages between them. It is therefore wise to select definite examples on the ground of the types it is proposed to adopt and to study each in detail. The selected examples should be carefully described and limits should be fixed to the variations from these typical descriptions, which will be admitted in stands to be included in the types. In this way a clear idea will be obtained of what is meant by each type, and it will then be possible by inspection to allocate all other stands in the forest to their places in the classification of forest types adopted. Any relationship between the types, as, for instance, their relative positions in a scale of ecological succession, should be explained, and their probable future development should be discussed.

It should not be forgotten that though the forester in Europe is often managing stands which are climax or sub-climax types for their sites, the forester in the tropics may often wish to manage a forest so as to maintain or produce a valuable sere which, if left to itself, would in that site progress to a climax with little or no economic value. The best known example of this is Sal (*Shorea robusta*) forest resulting from fire on sites which in the absence of fire would carry a climax of evergreen forest comprised of species which are not wanted. In this case, when the forester understands the position he can use fire as a deliberate means of achieving his object of management. A forester must study nature and endeavour to understand what nature is trying to do in a forest, not always with the idea of letting nature have her way, but sometimes in order to be able to guide and assist nature to produce a deliberately chosen result. The ecology of a great many tropical forests, particularly in Africa, has not yet been worked out fully, and until it is foresters will not be in a position to apply the best silvicultural treatment to achieve their objects of management.

The types selected should be given simple descriptive names, such as Scots Pine Forest, Oak /Ash High Forest, and will have to be plotted on to a Stock Map, which will then show the distribution of the various forest types or crops in the working plan area. In the case of a previously unmapped forest the boundaries of the types are recorded on each enumeration line as they are reached. When dealing with a forest which has already been divided into compartments, each compartment is examined separately and the types found in it mapped. For both methods it is necessary to have in mind a clear idea of each type crystallized by the effort made to describe selected samples in detail. Recording will then be reduced to noting the type by name with remarks about the variation of the particular stand from the standard of the type, e.g. 'Oak /Ash High Forest with some coppiced Beech now in top storey'.

It has already been suggested that a forest or crop type may have to be sub-divided into productivity classes because it is growing on more than one site, and its sub-divisions may need different silvicultural treatment, such as different rotations and thinning. In the case of pure even-aged crops for which yield tables are available the boundaries of the quality (productivity) classes of the tables can easily be mapped in accordance with the height for age differences in the crops. For mixed forest types, such as are often found in the tropics, some purely arbitrary and local productivity classes, such as a Class I in which the principal merchantable species attains a girth of ten feet or over at maturity without deterioration, a Class II in which few of this species exceed eight feet in girth without becoming hollow, and so on, are sometimes adopted. Differences of this sort are obviously due to factors of topography, climate, or soil, and can be used to complete the mapping of sites which has already been considered.

To be of real use to a forest manager a stock map, besides showing the forest types with their sub-division into productivity classes, should also show the distribution of age-classes. With even-aged stands this is simple: convenient age-class periods, 1–20, 21–40 years, etc., or size or treatment classes, such as seedling, thicket, pole, young, middle-aged, and mature timber, are decided upon to suit the particular case, and are indicated on the map, probably by various forms of hatching. A distinctive colour is often chosen for each forest type, and some foresters have used a range of light to dark shades of each colour to show the age-classes. As maps are made and used in all sorts of lights this practice is dangerous and in any case considerable skill is needed to keep each shade uniform. Safer methods are, either to use each type colour for hatching in different directions for each age-class, or to apply the colour solid and superimpose various forms of hatching in black. Boundaries of any productivity divisions of a single type can be shown by distinctive lines and the class symbols can be printed on the divisions. In Sweden, where each stand is numbered and managed individually, stock maps show the boundaries of each stand and information about each is printed in a standard formula on it. This is

$$\text{Number of Stand—Quality Class—Mixture of Species}$$
$$\text{Age Class—Cubic Metres per Hectare—Cutting Class,}[1]$$

and printed across a stand one may read such symbols as $\dfrac{9\text{–}6\text{–}820}{\text{VI–190–V}}$, meaning that stand No. 9 is of Quality Class 6, is composed of the species indicated by the figure 820 in an advanced age-class, carries 190m.³ to the hectare, and has been placed in a category of crops which should be felled and regenerated within the next ten years.

There are a number of different ways in which information can be presented on a stock map and a scheme should be selected which will be as clear and simple as possible, and suit the particular kind of forest which is being dealt with and the intensity of management to be undertaken. In irregular forest, for instance, the size-classes are all mixed up and cannot be mapped, but in some parts one size-class may be predominant and this can be indicated by a symbol printed on the appropriate portion of the map. The facts about the various forest types or the stands will be ascertained during the enumeration surveys carried out and noted in the compartment descriptions, from which such as are to be indicated on the stock map will be extracted.

Compartment Descriptions   It will be convenient to discuss now the Compartment Descriptions which form the most important Appendix to a working plan. The compartments being the permanent units of management and record require individual study and description. When dealing with a previously unmanaged forest, division into compartments will follow the investigation and survey of topography, soils, and forest types, and possibly even some consideration of the larger units of management which will be established. If the area has already been divided into compartments decisions must be made about possible alterations of these. An alteration of compartment boundaries is a serious matter which may break the continuity of long established records and cause confusion to executive staff familiar with the old boundaries. However, if the constitution is bad and the past records are not of much

[1] See page 134.

value, changes may be made after careful consideration. No area should be excised from a compartment simply because it needs different treatment from the rest now, if it is potentially capable of being treated in the same way as the rest later. In such a case it can be made into a Sub-Compartment temporarily, and designated by the compartment number followed by a letter. The sub-compartment then becomes the temporary unit of management, and what amounts to management by stands is possible without the disadvantage that stands, being only of temporary duration, are not convenient units for permanent records.

For simple descriptions of compartments, such as are required to enable future work to be planned, tabular forms are often convenient, and permit the data of several compartments to be put on one page. There is, however, another important function to be performed by the compartment descriptions. This is that each description should provide the basis of a permanent, running record of the history of a compartment, known as a Compartment Register or Compartment History, which ultimately requires a separate file for each compartment to which new sheets are added from time to time. It should be noted that long forms which have to be folded become tattered very quickly, and that for convenience of taking carbon copies the maximum size of sheet should be foolscap or preferably smaller, for ease of carrying in the field.

Bearing in mind these points it is possible to suggest a logical framework, within which the special requirements of any particular working plan area can be fitted. Such an arrangement is set out below and the following are the reasons on which it is based. For future convenience a brief, permanent description of each compartment individually, its area, situation, topography, soil, and site types, and past history, is recorded on the first sheet. As no advantage would be obtained from tabulation, this information is arranged in narrative form.

Then the temporary division into sub-compartments and the descriptions of their stands as they are now are set out on a second sheet. The selection of information to be recorded on this sheet depends primarily on the silvicultural systems and methods of yield control which are to be used. However, data which the officer who will revise the plan later will require to know should also be included. It must not be forgotten that an area which can only be managed very extensively now will probably be managed much more intensively later, and that one of the functions of a plan is to provide information on which such management will be based.

In the case of even-aged, pure stands, the important data are species, age, quality class, density, mean diameter, volume, and the operations to be carried out during the working plan period. For uneven-aged stands to be treated under the selection system the distribution of species and size classes, and the volume are required. In some continental plans only a short description of the stands, the percentages of each species present, the last operation, the total volume, and a graph of numbers of trees over size-classes, are provided after the permanent data of each compartment. In some tropical forests, where only certain species are likely to be useable, yield control refers to these only, and the numbers of them in different size classes and of their seedlings are the important data. This sheet is tabulated to avoid repetition of headings for each sub-compartment.

The recording function of the register is provided for by a third sheet to be maintained for each sub-compartment (stand) separately as long as it requires different management from the rest. On this sheet is recorded each operation as it is completed, with the results in produce of various kinds removed, revenue received, and cost, including non-remunerative tending and its cost.

At each revision only a new Sheet II will have to be prepared, the original being retained in the file as an historical document.

Sheet I remains unaltered and new pages of Sheet III only require to be inserted as the originals become full.

It should be noted that during the preparation of a working plan the column for Prescriptions for Future Operations (Col. (16) of Sheet II in the suggested arrangement) can be used for recording notes (in pencil) on what appears to need to be done in the various compartments. These notes, made at the time the compartments are examined and described, can then be considered in the light of the general scheme of management, the available funds and labour, and the relative urgency of the operations, and replaced by definite prescriptions of what shall be done in what years. The notes will save a lot of time which otherwise would have to be spent on going round to decide the relative urgency of, say, regeneration or thinning operations. If 'B Grade urgent' or 'soon' or 'can wait' has been written, only a few stands will have to be re-examined to decide their place in the thinning schedule.

In some large, understaffed forests the felling series, or the block, has to be adopted as the unit of record. Series, or Block, Registers can be kept on sheets similar to those used for Compartment Registers.

ARRANGEMENT OF A COMPARTMENT REGISTER

The headings shown are suggestions only and include various alternative forms of information which may need recording. Only such data should be recorded as are required,

(a) to be used in practical management according to the silvicultural systems, methods of yield control, and intensity of working applicable to a particular working plan area;

(b) to provide basic information about the permanent units of management, and to build up continuous records of the results obtained in them in order to assist the development of better management. It should not be forgotten that more intensive working is likely to become possible in the future, and that some data, which may not be of immediate practical use, are likely to be of value to future revising officers.

COMPARTMENT REGISTER SHEET I
PERMANENT DESCRIPTION

Compartment number
Area   Acres
    Productive   .   .   .
    Potentially productive   .
    Permanently unproductive .

          Total .   .   .

Boundaries
    North
    East
    South
    West
Site or Locality Factors
    Altitude
    Topographical Position
    Slope
    Aspect
    Exposure
    Geological Formations
    Soils, Drainage, surface
              soil
      Depth
      Fertility
    Site classifications,   Site ———   Acres ———
                        Site ———   Acres ———

History prior to the preparation of a working plan:

COMPARTMENT REGISTER SHEET II

COMPARTMENT NO. ....

DESCRIPTION OF GROWING STOCK AT .... (date)

Sub-compt.	Area, acres	Site Class	Forest Type and Percentages of Species	Mean Age or Size Class or Development Class or Size Classes present	Qual. Class	Mean Height total or timber ft.	Mean D.b.h. or G.b.h. ft. or in.	Density or No. per acre or No. per Size Class listed in (5)	Vol. per acre, cu. ft. or H. ft. o.b. or u.b.	Total Vol., cu. ft. or H. ft. o.b. or u.b.	C.A.I. per cent or Vol.	Description of Growing Stock	Description of Under-growth	Last Opera-tion	Prescrip-tions for Future Opera-tions
(1)	(2)	(3)	(4)	(5)	(6)	(7)	(8)	(9)	(10)	(11)	(12)	(13)	(14)	(15)	(16)

COMPARTMENT REGISTER SHEET III

COMPARTMENT NO. .... SUB-COMPARTMENT ....

HISTORY OF OPERATIONS TO DATE

| Year | Operation | Area Attempted acres | Area Completed acres | Species removed or established | Type of Produce removed | Quantity removed | | Value realized, £ s. d. | Cost, £ s. d. | Net Revenue, £ s. d. | Total cost per acre to date, £ s. d. | Total Net Revenue or Deficit per acre to date, £ s. d. | Remarks |
						Unit	Vol. or No.						
(1)	(2)	(3)	(4)	(5)	(6)	(7)	(8)	(9)	(10)	(11)	(12)	(13)	(14)

CHAPTER
8

Requirements of the Surrounding Population. Whenever a forest is brought under management the needs of the local people and the possibilities of local markets require first consideration. Even if the main object of management is protection or bulk supply of some form of produce to a distant manufacturing centre, it is essential to try to provide some benefit to the local population and to assure their co-operation.

In the case of a protection forest, required to safeguard a catchment which feeds a river on which some distant population depends, it must be remembered that restrictions are being applied locally which may make the management unpopular, and even lead, in extreme cases, to incendiarism or other forms of sabotage. Similarly people will resent seeing forest produce dispatched elsewhere if they have any difficulty in obtaining their own requirements. These facts are particularly true among primitive peoples with strong local tribal instincts.

In more highly developed communities the local market is more a matter of self-interest for the forest manager. The production of prime oak logs, for instance, will probably be quite uneconomic unless the inferior oak logs, the beech, hornbeam, sycamore, etc., that have to be grown with them, can be disposed of without long transport hauls. First consideration in forest planning has to be given to the satisfaction of any rights or customary privileges of the local people. Their orderly supply may necessitate some modification of local customs by consultation and arrangement with local bodies and the local government. On the other hand the existence of a right may preclude the adoption of some form of management under which the right could not be satisfied, such as the conversion of a beech forest, subject to rights of pannage into coniferous forest.

This chapter is not the place in which to state what shall be done, but is the place to record the facts and to indicate their bearing on the forest. In this first section only domestic and agricultural needs should be stated.

Markets and Marketable Products: Price Movements   In this section the working plans officer must collect all the information available about the possibilities of selling the various forms of forest produce which can be grown in the area, the present demand and present price for each, and the probable trends of future demands and of future prices. Study of past price fluctuations and their causes and of the possibility of improving marketing technique will be needed.

The silvicultural possibilities of the area have rightly been considered first and it is known what species can be grown. This chapter should settle the question of which of the possible species are the most economic to grow, so that in Part II prescriptions can be made without further argument. It is, of course, not only a question of species, but often of forms in which a species should be marketed. For instance, a high proportion of telegraph poles sold from a coniferous crop may compensate for a poor price for its timber, and so make it more worth growing than a species which yields a higher-priced timber but no poles. Again, where firewood is unsaleable owing to the distance from a market, it may be profitable to make and sell charcoal.

A working plan officer is often faced with one of two alternative situations,

    (*a*) too little production on his area to meet demand or

    (*b*) too small a demand to enable systematic management to be practised, and the forest thereby improved.

(*a*) This case justifies expenditure on any methods which will increase the output of the area, and reduce all forms of waste at any stage of growing, harvesting, or conversion. In the tropics it is liable to mean demand for certain primary species only, while wood of the more numerous secondary species is not considered to be timber at all. There is considerable scope for research into methods of making these species marketable through improved conversion, seasoning, preservative treatment, education of public opinion, and search for new uses.

(*b*) In order to attract a firm to establish a sawmill in a remote forest modifications of the ideal management will have to be permitted. It may be necessary to allow an economic volume of timber to be cut from a fairly small portion of the forest each year to economize transport. Thus, a long felling cycle instead of a silviculturally better short one may have to be adopted.

Lines of Export, Methods of Exploitation, and their Cost The system of internal transport existing in the forest tract should be studied to ascertain whether it is adequate or not. If it is not adequate, prescriptions for its improvement based on the conclusions reached in this chapter will have to be made in Part II. Here, and by suitable indication on the index map, the existing roads, tracks, rail- or tram-ways, rivers, slides, etc., and what portions of them are usable at what times of the year should be made clear.

The internal roads will, of course, lead into the export lines, whether rivers, railways, or roads, and the forester will have little influence on these. The market areas which can be reached by each and the costs per unit should be stated.

The existing agencies and methods for felling, hauling, loading, and extraction should be examined critically and their costs analysed. Any waste or possibility of improvement should be pointed out. If appropriate to the case in hand suitable sites for sawmills or other primary conversion plants should be located and discussed.

Staff and Labour Supply Here the existing staff, the distribution of work among them, the efficiency of the organization, or the need for reorganization should be stated. So should the availability of labour, rates of pay, standards, and costs of the various operations carried out. The possibility of expanding staff and labour should be discussed preparatory to the prescriptions which will be made for the numbers estimated as necessary to carry out the plan when the details have been settled. It is no use planning for expanding operations in areas where it is not possible to obtain the requisite labour.

Forest management, like any other form of management, depends for its success very much on the maintenance of good relations between the various grades of staff and the labour. Contented labour is economical labour and attention must be paid to conditions of work, with a view to their improvement if they are not satisfactory.

STATISTICS OF GROWING STOCK

The methods used for the collection of information about the volume of the growing stock and its rate of growth should be described fully enough for any forester reading the plan to judge how much reliance he can place on the accuracy of the data presented. The full data collected should be recorded in an appendix, but a summary sufficient to allow future proposals to be followed should be given in this chapter. This can take the form of a tabulated statement of species, volumes, age- or size-class distribution, and their rates of increment on the various sites. These data are required for the efficient handling of the crops and for the regulation of the yield. Yield is controlled by volume or by area, and in the latter case it is necessary to know the amounts of different kinds of produce which will be obtainable from the various areas.

Very careful consideration must be given to the organization of the collection of statistics of volume and growth, whatever methods of measurement and estimation are to be adopted. The unit of volume, cubic foot or Hoppus foot, under or over bark, in which the results will be expressed, must be selected and adhered to. The size at breast height of the smallest tree which will be measured must be fixed with care, and a suitable range of size classes adopted to suit the particular circumstances, not only with an eye to present needs, but also to those of the future. If for later stocktakings changes are found to be desirable, these will diminish the value of all past records. For instance, in one French forest the smallest tree measured in 1861 was over 7 inches quarter girth at breast height (i.e. one falling within the 8-inch q.g. class of a series of 2-inch classes), but in 1912 it was thought desirable to measure trees in the 6-inch q.g. class (5·1 to 7 inches q.g.). In all comparisons of the volume of the growing stocks of 1912 and later with the earlier records adjustments have to be made. In tropical forests sometimes only the species merchantable at the time have been enumerated, but this has generally been regretted later when markets were found for additional species. There was a case in India in which a demand arose for an unenumerated species after the enumeration had been carried out, but before the plan was completed, and some hurried additional work had to be done. Enumeration of growing stock is a fairly expensive and time-consuming business, and the inclusion in the work of all measurable trees of all species which grow with a timber-like habit to timber size will not increase the costs very much, and may save work and confusion later.

One hundred per cent enumerations, in which all trees are measured in suitable girth or diameter classes, and sufficient heights, total or timber, as required, are measured to enable reliable girth- or diameter-height curves to be prepared for each species and forest site, are the most reliable means of stocktaking, but are quite impossible to carry out in large forests.

Alternatively, sampling, either random or systematic, can be carried out at varying intensities, usually of from one to twenty per cent, according to the nature of the forest, the value of the stock, the degree of accuracy desired, and the time and funds available. When sampling units selected by chance are measured it is possible to calculate the standard error of the volume estimate. When strips or plots are chosen systematically it is not possible to calculate the standard error, but the greater convenience of the method in some circumstances compensates for not knowing the error, which is unlikely to be greater than that of random sampling of the same intensity and may even be less. It must be remembered that this standard error calculated by statisticians is the sampling error only. It does not include any errors caused by faulty measurement of trees, failure to measure all the trees in a plot, or measuring some trees which are outside the edge of a strip, or by using inaccurate volume tables to convert the measurements taken into volumes. Any one of these may be far greater than the sampling error, but often they receive less consideration, in spite of the fact that the practical forester in the field can do more to minimize them than the sampling error. The sampling (standard) error is likely to be decreased if sampling is stratified, that is to say if each forest type separately is divided up into sections within each of which the growing stock is more uniform as regards productivity than is the growing stock over the whole type, and if sampling units are selected, either at random or systematically, in each section separately. By this means either the standard error of sampling can be reduced for a given intensity of sampling, or a lower intensity of sampling can be adopted to give results within the limits of the standard error which is regarded as acceptable.

Thus, in regular forests of species for which yield tables are available standing volumes can be estimated for each age-class and quality (productivity)-class by determining the areas which fall within the quality class sites of the tables, their basal area per acre, and average density from samples. From the same data future increment can also be estimated. How much measurement is done will be a matter for decision in each case. A reasonable amount of sampling for each age and site section will give a reasonable estimate of the volume of the whole of that section, but not necessarily of the volume in any particular compartment. Unless the crops are very uniform an adequate set of samples within each compartment will mean considerably more work. In some cases it may be possible to concentrate on the older crops, particularly those to be regenerated during the working plan period, and merely to note the areas, quality classes, and approximate densities of the young and middle-aged crops. However, in intensive forestry the intermediate yields are often quite as important as the final yield, and estimates of production from thinnings classified by sales classes, e.g. telegraph poles, ladder poles, agricultural stakes, pitprops, and pulpwood, are needed. For the immediate future these can be based on sampling, either noting what should be cut out from the samples used for enumeration, which will save time but requires that the types of thinning to be made shall have been decided at the time of enumeration, or by the selection of special samples later. Yield tables and records of material taken out of similar stands at various stages of growth in the past can be used in the prediction of intermediate yields from the younger stands.

In unmapped irregular forests, particularly the extensive tropical ones, systematic strip sampling, combined with topographical, soil, and forest type survey, offers so many practical advantages that it is generally used. The trees are enumerated by species in convenient size classes, the records being closed at the end of regular, short lengths of line, and at any point where a change of forest conditions is noted. Local volume tables are prepared on the spot from which the numbers of trees can be converted into volumes. The principal forest types, and possibly different productivity zones, are mapped from the data collected during the enumeration survey, and it is therefore possible to combine records of volume applying to the sampling units of each separate forest type, or productivity site within that type, in order to determine the average volume per acre for that type or site. This average multiplied by the acreage of the type or site measured on the map gives its volume. In this way the principle of stratification is adopted when the survey results are compiled, although it could not be applied in the original sampling design.

In North America methods are being developed by which the volumes of coniferous growing stock can be estimated from aerial photographs. It is improbable that this will ever be possible in the case of mixed hardwood forests, but a great deal of assistance can be obtained from aerial photography in the differentiation of forest types and in the mapping of their boundaries.

Increment rates may be determined from yield tables, increment borings, or from periodic measurements of whole stands or of samples. The rates have to be found separately for different age-classes and productivity sites.

It must not be forgotten that growth may be offset to an appreciable extent by decay and by windfalls which cannot be salvaged, and that in the case of mature timber which has to be held over unfelled the volume which is harvested may be less than it is now. The full increment of a stand can only be realized if short cutting or thinning cycles can be applied.

Another point which a forest manager should remember is that what he wants to know is how much useable produce will be obtained from any area. Volume estimates by all the usual methods depending on measurements of basal area and heights assume that the whole of every tree will consist of millable logs or poles straight enough to use, etc. Allowance must always be made for logs which will be hollow, curved, or otherwise defective, and for poles of valuable dimensions which will only be fit for firewood. This can be very important, particularly in the tropics, if commitments to supply produce are

83

entered into on the strength of enumeration surveys. In one forest in India from which, according to the working plan, a certain number of logs capable of yielding two railway sleepers each were to be provided annually, many of the logs proved to be defective and only gave one sleeper each. The result was that in order to maintain the output of sleepers contracted for the forest was over-cut and the plan failed in its object. If a forest which is being organized, or a similar one on a similar site, has been worked, it should be possible to obtain figures from which percentage deductions for defective timber can be calculated and applied to the enumeration results to give the effective volume of each species.

Forest enumerations entail very arduous work, particularly in the tropics, and it has been found essential to pay particular attention to the conditions of employment of the staff and labour employed on them. Not only are special field allowances found to be worth while, but also the provision of suitable camping equipment and good rations. In this way experienced men can be attracted to and kept in enumeration parties and a high standard of accuracy obtained. Disgruntled staff whose only idea is to get off the job and who record any old figures, are a complete waste of money and provide a dangerous basis for planning. The working plan officer must himself check a proportion of the field work of all parties which he uses and discard unreliable workers, but he will never keep any reliable ones unless he sees that their conditions are as tolerable as they can be made, and their remuneration attractive.

CHAPTER
9

The area which is to be managed, its conditions and present state having been described in Part I, it is logical for Part II to open with a statement of what is required from the area.

Objects of Management These have been decided by the owner after he has considered technical advice, or in the case of state forests usually by the senior officer charged with the implementation of the Government's forestry policy after he has considered in consultation with his staff what part the area can play in the fulfilment of that policy. The management of forests can be directed to the attainment of a variety of objects, either singly, or, as is more usual, of several in combination. The main objects cover:

A. Protective Functions. These include:

(1) Preservation, rehabilitation, or improvement of vegetal cover on slopes to maintain or provide maximum absorption and storage of rainfall for steady release of water for consumption, irrigation, and production of hydro-electric power. Dense forest cover is one of the most effective and most permanent agents for the regulation of floods, which are liable to do extensive damage to farm lands and other property, the alleviation of the drying up of water supplies which is likely to alternate with floods, and the prevention of soil erosion, which silts up dams and navigation channels.

(2) Protection of villages, grazing land, roads, and railways in mountain valleys from avalanches of snow and rock, and from landslides.

(3) Amelioration of local climates by breaking the force of strong or desiccating winds, cooling the atmosphere, and raising the relative humidity locally.

(4) Protection of the soil on steep slopes from which it

would disappear under any form of land-use which involved the removal of the natural vegetation. In such cases the growing of some form of forest produce is the only economic use which can be made of the soil, and protection of the soil is the primary aim of management.

B. Productive Functions

Provision and maintenance of supplies of forest produce required for local agricultural, domestic, and industrial needs, and for export. Timber is often assumed to be the most important forest product, but this is not always the case. Firewood to replace cow-dung as fuel and so allow the latter to be used as manure is often of vital importance. Where poles and grass or leaves are the normal building materials the maintenance of high quality supplies of these is the most practical contribution which can be made to raising the local standard of living. Many minor forest products are essential to the life of the local communities, or are profitable articles of commerce (e.g. Christmas trees).

C. Amenity Functions

Maintenance of vegetation for appearance, recreation, or sport, including hunting, shooting, game photography, and fishing.

A famous example of the combination of protective and productive functions is that of the Landes of Gascony. Maritime pine was planted on shifting sand dunes and on unhealthy swamps on the landward side of the dunes caused by obstruction of water flow by blown sand. Some two and a half million acres of barren and unprofitable land were planted and by their protection transformed the agricultural and grazing value of the district. In addition they produce over two million tons of timber and twenty-seven million gallons of turpentine annually, which help to make the area one of the richest in France.

The mountain forests of Cyprus provide another illustration of a combination of protective and productive management. Left to themselves they tend to become even-aged and will only regenerate after considerable natural opening out, which allows erosion and floods to take place. Intensive group or strip working can overcome this and also provide timber which is very badly needed on the island.

Forest belts planted simply to stop erosion or avalanches will need management, often because the species planted originally will be unable to regenerate themselves and the succession will need assistance.

Amenity forests will become over-mature and deteriorate, so, in spite of public protests, they need skilled attention.

GENERAL PLAN

After having opened the chapter by setting out the objects of management the general organization by which these objects will be attained should be outlined. If different parts of the working plan area are to be devoted to different objects and will need to be managed according to different silvicultural systems, necessitating different sets of working prescriptions, these parts will conveniently form separate Working Circles. A working circle has already been defined (page 36) as just such a part of a working plan area. The definition also contains a rider to the effect that in certain circumstances working circles may overlap. A typical example of this is a teak-bearing forest in which the ten to twenty per cent of teak in the growing stock of much of the area is organized as a teak working circle, and the other merchantable timbers mixed with the teak and also found in the rest of the area are organized as a hardwood working circle.

There is a modern tendency towards greater flexibility in management, the growing of mixed crops and the adoption of stand management, which is leading to a less rigid conception of the term working circle. For instance, an area in which groups of conifers are mixed with hardwoods, hardwood regeneration is encouraged under coniferous crops and vice versa, may be treated as a single working circle in which each stand is given appropriate treatment as required. This calls for considerable skill on the part of the executive officer and requires an adequate staff of trained men. In a working plan for a forest of this nature the general objectives, such as the types of produce to be aimed at, the species whose regeneration is to be encouraged at various stages of the development of the stands, etc., are laid down. Prescriptions for operations in sub-compartments are made for short working plan periods, and latitude is given to the executive officer to vary the timing

87

and intensity of their application to suit the progress of the stands, the incidence of good seed years, and current market demands.

The formation of different working circles for broadly different types of forest still remains a valuable means of simplifying management and does not prevent reasonable diversity of stock and flexibility of management.

Division of the Area   This cannot, of course, be settled until all the decisions have been made about what species are to be grown under what silvicultural systems and on what rotations, though these decisions are recorded later in the plan.

As compartments are the smallest permanent units of management they must be the smallest unit of allocation to working circles. The only exception to this is when a Conversion Circle is formed with the object of converting certain areas from one type of forest to another, say, coppice to high forest, hardwood to conifers, etc. A conversion circle is a temporary management unit and can contain some of the smallest temporary management units, that is to say, sub-compartments, when these are to be converted to conform with the rest of the growing stock of their compartments.

When intensive forestry is possible stand management can be practised in the manner just described, but normally in large areas a certain amount of detailed attention must be sacrificed in the interests of practical and economical management. If an entire compartment is not suitable for inclusion in one working circle it is not a suitable compartment and should be split. However, there is a limit to what is practical, which must be determined by the circumstances of each case. A private owner with markets on his doorstep for every beanstick he can cut in his small forest may find it possible and profitable to manage it on the lines of a market garden, but a tropical forester with several thousand square miles of jungle to manage with a few semi-trained subordinates and a market a hundred miles away, will have to think in terms of large areas.

Examples of working circles which may be formed, in addition to a Conversion Working Circle already mentioned, are:

In a forest, part of which is to be worked as high forest for timber and part as coppice for fuel:

A Timber (or High Forest) W.C.

A Fuel (or Coppice) W.C.

In a forest in which good sites are to be used for hardwood timber production and the rest for coniferous timber:

A Hardwood W.C.

A Coniferous W.C.

In both cases if certain parts, say, round the owner's house or a beauty spot of public resort, had to be managed in a different way from the rest, possibly under a selection instead of a uniform system, for the sake of appearance, an Amenity W.C. might be formed.

When a forest has been divided into compartments the allocation of these to their appropriate working circles should be fairly easy, but there will be border-line cases. In deciding them, such matters as the desirable size of the different circles, the geographical situation of a compartment in relation to others in the possible circles should be given consideration. Working circles need not be continuous areas, but may consist of a class of site or forest type scattered in separate compartments throughout the working plan area, though it is certainly convenient if they can be compact blocks of compartments.

If the working plan area is one which has not previously been divided into compartments, it is possible that the division into circles will be made before that into compartments. The map built up from the grid survey and enumeration on which the distribution of site and crop types was shown would be the basis for decision about the distribution of areas to their appropriate circles, and then compartments would be formed of sizes suitable for the intensity of management intended for each circle.

Site is more important that crop type in allocation. Different stands on the same site can ultimately be converted to a similar type, but differing sites are almost certain to produce different results. For example, in a working plan area in which a Hardwood W.C. and a Coniferous W.C. are being formed there may be some coniferous stands in a block of hardwood forest. These may have come into existence through planting, or as a result of fire followed by the invasion of coniferous seed from outside. In such a case it would be quite justifiable to incorporate them in the Hardwood W.C. and to arrange for their conversion back to hardwood at the most suitable times, taking into consideration both the most profitable stage to dispose of their produce and the future hardwood age-class distribution which is desired. In this case it would not matter whether the coniferous stands had been constituted as compartments or as sub-compartments in hardwood compartments. If, however, these coniferous stands were due to a site difference, say, the existence of dry, sandy areas in what was otherwise a loam, then they ought to form part of the Coniferous W.C., because they will grow fair conifers, but only poor hardwoods. In this case, the question of size comes in, and only if they are large enough to form compartments can they be placed in the Coniferous W.C. This, however, should not prevent them being managed as coniferous groups or belts in the Hardwood W.C. Management must not be so inflexible as to prescribe faulty silviculture for the sake of regularity or convenience of recording. The point, however, is that if two sub-compartments of the same compartment are managed under different sets of prescriptions to attain different objects, they are really two compartments whatever you call them. When the allocation to the working circles which it has been decided to constitute has been completed a brief summary of the principles followed in arriving at the decisions should be given, and a table made showing the constitution of the circles which account for the total acreage of the working plan area. It is not necessary to give here lists of the compartment forming the various circles, as these will be needed in the separate chapter for each working circle. In Part II of a working plan there should be no arguments, but statements of decisions which follow from the facts described in Part I on the lines of 'because such and such is the case, so and so shall be done'.

The prescriptions themselves should be definite, concise, and clear, and should not include any reasons: 'There shall be two Working Circles, a Hardwood Working Circle and a Coniferous Working Circle.

(1) The Hardwood Working Circle, of . . . acres, comprises all the areas of brown forest soil with free drainage, including some . . . acres at present carrying Douglas Fir . . .'

90

Compartments If there is anything to be said about the division of the areas into compartments, e.g. the principles which have been followed in forming them, their boundaries or the method of their demarcation, it should be stated here. If the plan is one for an area already divided into compartments the reasons for any changes made should be stated.

Period of the Plan This has been discussed on page 47. The period which should be stated here is that for which the detailed prescriptions of the plan will apply, and should be given as, for instance, 'The ten-year period from 1st January, 1950, to 31st December, 1959.' The date on which the revision of the plan is to be started, e.g. one year before the expiration of the present plan, should be prescribed.

If the working plan area has been divided into working circles it will now be necessary to set out the plan of management for each circle separately, so a separate chapter for each circle should be written. If the working plan area is to be a single working circle the sections listed below under the chapter heading 'Special Plan', or such of them as are needed in the type of plan being drawn up, should be included in Chapter I of Part II as additional sections, otherwise Part II of the plan will continue with,

Chapter II. Special Plan for the (Mahogany) Working Circle.

Chapter III. Special Plan for the (Ironwood) Working Circle, etc.

Constitution of the Working Circle This is best given as a table setting out what forests, blocks, and compartments are included in the circle, with the area of each.

Sub-division of the Working Circle It has already been pointed out that a working circle may be divided into two or more felling series, which are management units with separately regulated yields, and that a felling series may be divided into two or more cutting sections in order to regulate the cuttings in some special manner (see Fig. 5). The area of the working circle, its distribution, topography, and climate, its stocking, the rotations, silvicultural system, and intensity of working proposed, the markets to be served, and the distribution of available labour, determine whether or no division into felling series is required, and what their number and extent should be. If the working circle is a large, undivided block, intensively worked, the annual yield from one regeneration area might be larger than is convenient to handle as a unit, and its harvesting from one annual coupe might expose or open a dangerously large area in one place. If the circle consists of a number of scattered forests, or covers both slopes of a ridge, it may not be convenient to produce all of each year's major produce from one part. For instance, in mountain valleys the produce of the valley slopes will have to be extracted downhill into each valley, and probably processed in a mill in each. Each of these sawmills will want a sustained yield of timber, and on the slopes above each a normal series of age- or size-classes should be built up and a permissable yield fixed. In other words, a separate felling series should be formed. This assumes that the whole collection of ridges is going to be worked with the same object of management under the same silvicultural system for the production of sawn timber, that is to say, it will be one working circle.

The present distribution of age-classes does not affect the division into felling series, in each of which a normal growing stock with its own normal yield will be built up as soon as possible. The beginnings of separate series of age- or size-classes may well be adopted as the basis of felling series, if they are convenient.

When forestry is intensive, as on the Continent of Europe, working circles are generally divided into several felling series of quite small size. When it is extensive, as in tropical colonies, each circle is commonly organized as one felling series, unless it is possible to obtain from it more than one annual economic supply for a sawmilling concession.

To sum up, working circles are divided into felling series,
(1) when several markets are to be served,
(2) when topography or geographical distribution enforce the use of more than one main extraction system,
(3) to avoid having too large an area in any one place under regeneration or other treatment,
(4) to distribute working or the labour force over the circle, either as a protective measure against outbreaks of fire, etc., or to economize in travelling time for labour.

Felling series are formed for management reasons, and there

may be silvicultural reasons which make their division into cutting sections desirable. In regular forests it may be that dangers from fire, wind, or insects make it desirable to break up the sequence of age-classes. In irregular forests managed under the selection system when the whole area of a series cannot be worked over every year, it is divided into as many cutting sections as there are years in the felling cycle.

The allocation of compartments to felling series and cutting sections, if any, is best set out in a table.

Species, Silvicultural system, rotation or exploitable size.   In this section are recorded the decisions which have been reached about these essential points of management. The reasons should be stated briefly by reference to facts and figures recorded in Part I, and without argument.

The deciding factors are the maintenance of soil fertility and of healthy crops suitable to the sites and capable of being regenerated thereon. The crops grown must be such as will satisfy the objects of management of the owner and, in the case of production forests, must provide the types and qualities of produce which are required for particular purposes, or which will be the most profitable to grow without harming the sites.

The proposed distribution of species to sites, or the proportion of species to be aimed at in the case of mixed forests, the rotations or provisional rotations, or exploitable sizes should be stated.

The silvicultural system must be based on the silvicultural requirements of the crop to be grown and on the local conditions. If the system is a well-known one it can be prescribed by name only, e.g. clear-felling with artificial regeneration by planting. There are, however, more systems of silviculture than have been described or named. Not only can there be combinations of different types of harvesting leaving different conditions for regeneration, and different methods of regenerating under these varying conditions, but also combinations, such as supplementing natural regeneration by planting or sowing the same or other species. It may, therefore, be necessary to give a short description of the system to be used. In any case very clear and precise prescriptions for all the silvicultural operations required must be drafted. These prescriptions will often have to be carried out by men without any professional training in forestry, and one of the benefits of a working plan should be that it enables essential work to be done where and when it is needed by foremen or other subordinate staff.

CHAPTER
10

The layout of this section will depend on the method of regulating the yield which is adopted. Broadly, yield regulation consists of estimating what the French call the Possibility, i.e. the productive capacity of an area, deciding how much of this shall be removed, how much reinvested in the wood capital of the area, or how much of any excess wood capital shall be removed in addition, and how and from what portions of the growing stock the cutting shall be done.

In a normal forest the whole annual increment could be cut each year and the same capital would always be left. Obviously, if the growing stock is not normal but consists of young trees only, the wood capital is too small and the increment must be left uncut to build it up. If the bulk of the growing stock is old there will be too much wood capital on the ground and the incremental rate will be low. The capital will need to be reduced by cutting more than the increment to make room for more young trees. In a production forest the object of a forest manager is usually to obtain from his area a fairly steady yield, while he is endeavouring to bring it into a state in which it will give a sustained yield equal to the maximum obtainable from the correct wood capital invested in the soil under the climatic conditions present.

It has already been pointed out (page 33) that most methods of calculating the possibility of an area were devised to suit some particular forest worked under a particular silvicultural system. Some of these methods can only be used in conjunction with that system, but others have been borrowed and used, with or without adaptation, in different types of forest worked under different silvicultural systems. To make any attempt to classify the methods according to silvicultural systems would be confusing, and it is better to classify them according to the principles on which they are based. Some methods calculate a final yield, which is the yield from regeneration fellings only, and some a total yield which includes all thinnings and tending fellings as well (see page 24).

A forester faced with the organization of a forest in Africa or Malaya can study what others have done in Europe and India and see how far he can make use of the principles they have applied to the regulation of yields. He cannot expect to find that all the details of any of their methods will suit his special requirements, any more than that all the details of their silvicultural systems will suit his crops, soils, and climate.

METHODS OF CALCULATING THE POSSIBILITY OF A FOREST

The main principles on which methods of calculation have been based are given in the first column, but only a few examples of particular methods making use of the various principles are listed in the second column.

METHODS BASED ON	EXAMPLES OF PARTICULAR METHODS
A. Area only.	A1. Annual Coupe Method (*Méthode à tire et aire*).
B. Area and Volume.	B1. Permanent Periodic Block Method (*Méthode à affectations permanentes*).
	B2. Revocable Periodic Block Method (*Méthode à affectations revocables*).
	B3. Single Periodic Block Method (*Méthode à affectation unique*).
	B4. Judeich's Stand Management Method (Judeich's *Bestandwirtschaft*).
C. Volume and Increment of whole Growing Stock:	
(*a*) Actual present Volume only.	C1. Von Mantel's Method.
	C2. Symthies' Modification.
(*b*) Comparison of actual and normal volumes.	C3. Austrian Method.
	C4. Heyer's Modification.
(*c*) Comparison of successive enumerations and volumes removed.	C5. Biolley's Control Method (*Méthode du contrôle*).
(*d*) Size Classes.	C6. French Method of 1883.
	C7. Regeneration Area Method (*Méthode à quartier de régénération*).
	C8. Brandis's Method.
(*e*) Development, Tending, or Cutting Classes.	C9. Swedish Method.

This method is the oldest form of regulating the cutting of a forest. It only regulates the final yield, that is to say the cutting of trees in the areas to be regenerated, and does not take into account production from thinnings and other operations in the younger stands, though separate arrangements can be made for this to be fairly uniform each year.

In the 14th century the need for some restriction of cutting was realized in France and regulations were made under which areas to be cut over each year had to adjoin each other (*couper à tire*) and be of not more than a certain area (*couper à aire*), hence the name *méthode à tire et aire*. It was first applied to coppice crops worked on rotations of up to twenty years, and the area of the annual coupe was restricted to the number of acres in the felling series divided by the number of years in the rotation. Thus, in the case of a twenty-year rotation one-twentieth of the available forest was cut as an annual coupe each year in regular succession, the first coupe being cut again in the twenty-first year. Later the silvicultural system of coppice with standards was introduced with the same method of area control, a certain number of trees being reserved for standards of one, two, and three times coppice rotation age each time a coupe was entered. In the high forests, particularly those of oak which it was desired to regenerate by seedlings, attempts were made to apply this annual coupe method, leaving some five to ten seedbearers per acre in each coupe in the hope of obtaining natural seedling regeneration which could be felled at the end of a hundred-year rotation.

It is interesting to note that this area regulation method is being used to-day in Nigeria, where a provisional rotation of a hundred years has been adopted and the concession area allotted to each timber firm has been divided into a hundred equal-sized annual coupes, one of which is worked over for timber each year. The silvicultural system used, called the Tropical Shelterwood System, is different, but the regulation of the yield is the same, i.e. the average annual yield is the average volume of timber found on one-hundredth of the area of a felling series. Naturally, in mixed tropical forest this volume varies considerably from year to year, particularly in

[1] See also pages 173, 180.

regard to the amounts of particular species.

In Europe at present regulation by area is only applied to fairly regular coniferous forests growing in conditions favourable to regeneration, for instance, the Maritime Pine forests of the Landes of Gascony. The annual coupes need not be adjoining (*à tire*), but the old name of *à tire et aire* is still used.

In the tropics the method is used fairly generally for plantations worked on short rotations, particularly those coppiced for fuel. The example used for a theoretical normal forest in Fig. 1 illustrates the principle. There are as many annual coupes as there are years in the rotation, and the final yield obtained each year is the volume grown on the year's coupe during a rotation. If the annual coupes are equal in area the annual yields will only be equal if each coupe is of equal productivity. If the coupes are not of equal productivity the annual yields will vary, but over a rotation the average annual yield will be equal to the sum of the increments put on by each age gradation in one year. In the case of a short rotation this equalization of the yield over a short period may be satisfactory. If it is not the annual yields can be made more regular by allocating to each coupe an equi-productive area instead of an equal area. This can be done if the various site (or productivity) areas in the felling series are mapped and measured, and if the yield per acre at rotation age of each site is ascertained. If Yield Tables are available these will give the average volumes for several different productivity sites, and also the heights at various ages of trees growing on these sites, so that by measuring the heights of the various age-gradations it is possible to map the sites. The areas of the sites can then be measured and expressed in terms of a common productivity according to their productive capacity at maturity. For instance, with an eighty-year rotation yield tables for a species might show that:

Qual. I sites produce 6,000 H. ft. per acre at 80 years

,,   II   ,,   ,,   5,400   ,,   ,,   ,,   ,,

,,   III   ,,   ,,   4,200   ,,   ,,   ,,   ,,

Then 1 acre of Qual. II site $= \dfrac{5,400}{6,000} = 0.9$ acre of Qual. I site

1   ,,   ,,   III ,, $= \dfrac{4,200}{6,000} = 0.7$,,   ,,   ,,

In order to obtain a normal series of age gradations it would be necessary to increase the area of those age gradations which were on the poorer sites. If the annual coupes on Q.I sites were 10 acres each those on Q.II would have to be 11·1 acres and those on Q.III sites 14·3 acres each in order to obtain equal annual yields. Often, of course, the age gradations would be partly on one site and partly on another. The usual method of organizing a series for area control is to site-map the whole, express the total area in terms of the commonest site, and lay out each annual coupe to be equivalent to an equal number of acres in terms of the common site.

Thus, using the production figures given above

Site.	Acres.	Factor to express as Q.II.	Acres equivalent to Q.II.
I	50	$\dfrac{6,000}{5,400} = 1.1$	55
II	525	1.0	525
III	425	$\dfrac{4,200}{5,400} = 0.8$	340
	1,000		920

The average annual coupe would still be $\dfrac{1,000}{80} = 12\frac{1}{2}$ actual acres, but each would be laid out to contain the equivalent of $\dfrac{920}{80} = 11\frac{1}{2}$ acres of land of Q.II site, and they would vary from 10·35 acres of Q.I ($11·5 \times \dfrac{5,400}{6,000}$) to 14·8 of Qual. III ($11·5 \times \dfrac{5,400}{4,200}$) or any combination, such as 5 acres of Q.II and 8·3 acres of Q.III.

The area equivalent to any particular site is usually referred to as a Reduced Area, and the factors to calculate it as Reducing Factors, but it is not always reduction, and the term is really more appropriate to calculating the reduced volume of an understocked area by comparison with the volume per acre of a fully stocked one. It should be noted that even when the annual coupes have been laid out to contain potentially equi-productive areas, the yields obtained from each will only be equal if they are equally stocked. Yield tables are prepared from figures of what are assumed to be fully stocked stands, and in practice most stands will not carry full stocking. This variation caused by stocking cannot be dealt with in the same sort of way as that caused by differences of site quality, because the stocking factor is not a permanent one, but will vary from age to age and from rotation to rotation. It must, however, be borne in mind when estimating the volume which will be obtained from a given size of annual coupe. Density, which is taken as unity for a fully stocked area and expressed in decimals for partly stocked woods, is usually assessed in terms of the proportion which the actual basal area of trees per acre bears to the yield table basal area per acre. Thus, if b.a.p.a. of a stand is only 160 H. sq. ft. instead of the 200 shown for its age and quality in the tables, its stocking is

$\dfrac{160}{200} = 0\cdot 8$. If the volume per acre shown in the tables was

6,000 H. ft. all that could be harvested would be $6{,}000 \times 0\cdot 8$ $= 4{,}800$ H. ft.

The object of yield regulation is to make available fairly constant supplies of produce for local requirements, or as the raw material for processing plants such as saw-mills or paper factories. Steady provision of raw material for current needs or to keep machines and labour employed will be needed, but demands fluctuate with market conditions, and a forest manager must be prepared for a certain variation in demand from year to year, which, however, will probably average out over a period of years. It is, therefore, not worth while trying to organize absolutely equal annual yields, and unless the variations in stocking are considerable it will be sufficient to find the average stocking of, say, the next ten annual coupes and from it predict the average annual yield for that period. In any case the stocking of the rest may well alter in ten years. If, for instance, in the case of the felling series cited above, the average density of the first ten coupes was $0\cdot 75$, the average annual yield which could be expected would be:

Area of Coupe in terms of Q.II site $\times$ Final Yield per acre of Q.II $\times$ Average Density, or $11\frac{1}{2}$ acres $\times$ 5,400 H. ft. $\times$ 0·75 $= 46{,}575$ H. ft., and 46,500 H. ft. should be a fairly safe annual estimate. There will be variations, but 465,000 H. ft.

should be available in the ten-year period. If there were wide variations of stocking, adjustments in the size of temporary annual coupes might have to be made, but in such a case the area method of yield regulation would be unsuitable, and some other method would probably be easier to apply.

In unmanaged forest it is not practicable to lay out equi-productive annual coupes, but in Nigeria some measure of avoidance of large fluctuations is sought by leaving a per-centage of a concession area unallocated to annual coupes. For instance, for a concession area of 110 square miles annual coupes of one square mile for a hundred-year rotation may be prescribed, thus leaving ten square miles which can be used to replace or supplement very poor coupes.

The area method is a very crude one of regulating yield in mixed natural forests, and was given up in Europe early in the 19th century in favour of regulation by area and volume combined, which is described in the next chapter. It did, how-ever, render valuable service in bringing order out of chaos, and in similar primitive conditions (e.g. in parts of the tro-pics) still has its uses.

CHAPTER

II

B1. REGULATION BY AREA AND VOLUME COMBINED
This method is also confined to an estimation of the Final
Yield from areas to be regenerated. As in the case of the Area
Method thinnings in the younger stands are usually carried
out on a fixed thinning cycle or cycles (say, a three-year cycle
for crops under a certain age and a six-year cycle for those
over). In this way a regular area is thinned each year and the
volume of thinnings obtained will be fairly regular, so can be
estimated and added to the final yield to give a total yield.

The area and volume combined methods were a direct development from *tire et aire*, because it was found that annual
coupes in high forest could not always be regenerated regularly, and that for many species natural regeneration was best
secured by a gradual reduction of the canopy of the mature
trees.[1] A clean forest floor under full canopy was required as
a seed bed, and a light thinning, called a seeding felling, let in
enough light for seeds to germinate without a lot of weeds,
which would have resulted from a heavy felling. Once obtained, the regeneration needed progressive increases of light
by secondary fellings, and eventually complete use of the site
after a final felling. This system, which was first devised in
Prussia over two hundred years ago, meant that regeneration
was carried out over a period which varied with the species
and locality, was selected to suit the local conditions and to
be a convenient fraction of the rotation.

Suppose that fifteen years was selected as a convenient time
in which to complete the regeneration of a mature stand in a
particular forest. Then, in theory fifteen of the old annual
coupes could be grouped together into a Periodic Coupe or
Block which could be felled and regenerated during this period
of fifteen years. Instead of cutting the mature trees in succes-

[1] See also page 178.

sive coupes by area, seeding fellings would first be made over the whole block, followed by secondary and final fellings. If the rotation was ninety years there would have to be six periods of fifteen years each during each of which a similar block was felled and regenerated.

Actually, of course, the old annual coupes had nothing to do with fixing the areas of the blocks, except that because the coupes had been worked in order, it was possible to form compartments containing approximately equal aged crops, and to group the compartments into fairly equi-productive blocks to be regenerated in turn. The area of each block in a uniformly productive felling series should be the area of the series divided by the number of periods into which the rotation is split. The number of the periods is the rotation divided by the regeneration period and, of course, there would be a corresponding number of blocks, one for each period. The blocks were numbered with roman numerals, P.B.I. consisting of the oldest mature crops to be felled in the first period, P.B.II of the next oldest, and so on down to the last block to be regenerated in the last period of the rotation and formed of the youngest crops in the forest.

Thus, in a forest of 2,000 acres worked on a hundred-year rotation in a locality where a mature stand can be regenerated in twenty years, there would be five $(\frac{100}{20})$ regeneration periods of twenty years each, during each of which a periodic block of 400 $(\frac{2,000}{5})$ acres would be regenerated. If adjustments of the actual areas of the blocks should be necessary to make them approximately equi-productive, these could be made according to the productivity of the sites as described in the area method of yield control.

The area of P.B.I. as selected regulates the final yield to be cut in the first period, but as seeding fellings have to be carried out all over the block it is necessary to find out approximately what volume can be cut each year, in order to assure fairly even annual outputs and the completion of felling by the end of the period. This is done by measuring the volume of the growing stock standing in the block at the beginning of the period, usually by a hundred per cent enumeration on the Continent, and by sampling where less intensive forestry is practised. As the last tree will not be cut till the end of the period, the average time which the trees will stand and put on increment is half the period. Therefore the increment which the stand will put on in half the period must be ascertained by one of the usual methods, increment borings, stem analyses, yield tables, etc., and added to the original volume of the stand to find the total quantity which has to be cut in the period. This volume divided by the number of years in the period is the maximum annual yield.

Thus, if the 400-acre P.B.I. mentioned above had a growing stock at the beginning of the first period of 1,600,000 H. ft. and its increment rate for the next ten years, i.e. half the regeneration period of twenty years, is estimated to be one per cent per annum, the permissible annual yield would be calculated as follows:

Present Growing Stock measured . . . 1,600,000 H. ft.
Estimated increment on this for 10 years . . 160,000 „
Total volume to be cut in the 20-year period . 1,760,000 „
Maximum annual yield . . . . 88,000 „

The principle was expressed in a simple formula by Cotta, the founder of the first forest school in Saxony,

$$\text{Annual Possibility} = \frac{V}{p} + \tfrac{1}{2}Vt$$

when V = measured volume of growing stock in a Periodic Block.

p = number of years in the regeneration period.

t = the annual increment per unit (e.g. 0·01, which is 1 per cent of 1).

using the same figures A.P. $= \dfrac{1,600,000}{20} + \tfrac{1}{2}(1,600,000 \times 0 \cdot 01)$

$= 80,000 + 8,000 = 88,000$ H. ft.

In addition to this final yield there will be production from thinnings in the other periodic blocks. These thinnings are carried out entirely in accordance with the silvicultural needs of the various stands, but they naturally tend to provide approximately similar amounts of material year by year, so after a time the average annual yield of thinnings can be estimated.

The method just described is that of *Permanent Periodic Blocks*.[1] The intention was that each block should be felled and re-

[1] See page 190.

generated in turn in the period to which it was allocated, and that at the end of one rotation the whole series of blocks would form a completely normal series of age-classes, which could be worked round in the same order for ever. In making the original allocations to blocks certain compromises were inevitable in order to secure regularity. For instance, a compartment which might have been regenerated in the first period was held back till the second in order to balance the blocks, or another was felled too young, but such sacrifices were considered a small price to pay in order to produce a normal series in one rotation.

However, living woods cannot be treated as machines: regeneration did not always appear or develop to order, and final fellings had to be postponed in some compartments: storms blew down trees and started regeneration in compartments of blocks not due for regeneration: fire, insects, and wars caused unforeseen complications.

The method has therefore been abandoned, except for just a few hardwood forests in the plains of France, where the climate is favourable for regular seed years and even development of regeneration, and where storms are rare. It is also still used in one of the coniferous forests of the Jura, but with many safeguards. Even in these cases the retention is largely for sentimental reasons, and is often more nominal than real. The forest of Belleme, in Normandy, is one of them, but now, in the fourth period, in one of the series a compartment of P.B.II is being regenerated, and one of the compartments of P.B.IV was regenerated in an earlier period. In all but name the method being used now is a modification of the Permanent Periodic Block method which was the first modification to be made, and was called

B2. REVOCABLE PERIODIC BLOCK METHOD[1]

In this method the working plan officer who revised the plan at the end of a period was allowed to rearrange the compartments and even sub-compartments into a different set of periodic blocks, without any real intention that these shall be permanent. Compartments in which regeneration had been started, whether intentionally in the previous period, or acci-

[1] See page 199.

dentally by storms, but not completed, are the first to be allocated to the current regeneration block. As their stocking has already been reduced by seeding fellings or windfalls, their effective area to count towards the total of the blocks has to be reduced in proportion. It must be remembered that some of these compartments will have been included in the previous regeneration block, and that if the actual area of the current block is not larger than the total area of the felling series divided by the regeneration period, the whole forest will not be worked over in a rotation. It may be that as little as a quarter of the volume of a normally stocked stand is left in a partially regenerated compartment, and hence only a quarter of its area should count towards the normal area of the current block. Allocation of potentially equi-productive blocks was not important, as the blocks were to be temporary only. The method of fixing the area is dealt with in more detail in the next section, where it is even more important. Any idea that the compartments for regeneration had to be all together in a self-contained block was abandoned, and single compartments or sub-compartments scattered all over the series could be selected and constituted as parts of blocks. Indeed, the fact that the regeneration fellings were not all adjoining came to be cited as one of the advantages of the method, which is still used in most of the hardwood forests of France. In this method the rotation is of less importance and less rigid than in the method of permanent blocks. It would, in fact, be possible to regenerate some compartments twice in a rotation, say, in the first and last periods, and some not at all. The next modification was

B3. SINGLE PERIODIC BLOCK METHOD[1]
It is fairly obvious that, if the constitution of all periodic blocks is going to be revised at the end of every period, the only really important one is the current block, in which all the main fellings will be done, and that only this one need be constituted. However, if this is done it is very important that its area shall not be too large, in which case the forest will be overcut, or too small, in which case the whole forest will not be regenerated in a rotation.

[1] See page 201.

As before, a regeneration period is fixed of such a duration that natural regeneration can be obtained as a sequel to gradual exploitation. The normal area 's' which could be allocated to the first period is calculated as

$$\frac{\text{Number of Years in Regen. Period}}{\text{Number of Years in Rotation}} \times \text{Area of Series}$$

Those compartments in which regeneration has been started, either in the course of regeneration operations in the last period or by wind, but not completed, are picked out first. Let their full surface or ground area be called 'e'. In order to determine their normal area in terms of forest in which regeneration has not been started the basal area of all trees in this area 'e' is measured. Let the actual basal area be called 'a''. Then the normal basal area 'a', which complete stands covering the area 'e' would have, is calculated according to the average basal area per acre found in mature compartments in which regeneration has not been started. It should be noted that hundred per cent enumerations are usually carried out over the whole block to be regenerated, so these partially regenerated compartments and a number of compartments ready for regeneration would be measured, and this work would provide the data required. The reduced or normal area 'e''' of the incomplete stands can then be found because $e' : e :: a' : a$, whence $e' = \dfrac{a'}{a} \times e$. The balance between the normal area 's' to be allocated to the regeneration block and the reduced area 'e''' of the partial stands already placed in the regeneration block is then made up by selecting compartments which are ready for exploitation and regeneration. An example will make this clear.

If Total Area of Felling Series = 2,000 acres
   Rotation   =   100 years
   Regeneration Period   =   20 ,,

then Normal Area of Block = $\dfrac{20 \times 2,000}{100}$ = 400 acres   s

If Actual Area of Compartments partially opened =   300 acres   e'
  Actual Basal Area of Trees in these Compartments = 20,000 sq. ft.   a'
  Average Basal Area per acre of unopened stands =   200   ,,
  then Normal B.A. of the 300 acres partially opened = 60,000   ,,   a

and Normal Surface Area of this 300 acres $= 300 \times \dfrac{20,000}{60,000}$

$$= 100 \text{ acres}$$

The area of untouched compartments to be added to the Regeneration Block to complete it $= 400 - 100 = 300$ acres.

(The total surface area of the Regeneration Block will thus be 300 acres partially regenerated and 300 acres not yet touched, total 600 acres.)

The annual yield from the regeneration block is calculated as before by dividing the measured volume standing on it, plus the estimated increment on this volume for half the regeneration period, by the number of years in the period.

The rest of the forest or series is divided into one or more blocks of thinning coupes which are thinned on selected thinning cycles. In the example given above there would be 1,400 acres in the thinning block, and on a ten-year cycle 140 acres would be thinned each year. Or there might be two thinning blocks, one consisting of the compartments up to a certain age, and the other of those above that age, with different thinning regimes and cycles prescribed for each. The production from these thinnings would, of course, be additional to the final annual yield from the regeneration block.

It should be noted that this method is really only suitable for fairly regular, evenly stocked forests, such as have been produced in Europe by long management. If there is much variation in stocking, the method of estimating the normal area of stands in which regeneration has been started would be dangerous. As in the method of revocable blocks the rotation need not always be adhered to strictly, and any compartment which would benefit by being felled and regenerated can be so treated in any period, while stands of rotation age which are still putting on good increment can be left.

The method is sometimes called *Méthode à affectation unique mobile* in France, whence the name Floating Periodic Block was derived, but the terms *Affectation Unique* and Single Periodic Block are better. In France the compartments for regeneration are always coloured blue on the management map, and sometimes the name *Quartier Blue*, Blue Area, has been applied loosely to the single periodic block. The name actually belongs to an entirely different method in which regulation of the yield is by volume (see page 128), and should not be used for a periodic block method. *Affectation Blue*, Blue Periodic Block, would be more appropriate for this single regeneration block.

There was a modification of this method in which the area for regeneration was not strictly a periodic block at all. In this the area of the regeneration block was fixed by picking out all partially regenerated compartments, reducing their area to normal in the manner explained above, and adding the full surface area of any other compartments which were ready for regeneration. The regeneration period, or rather the exploitation period, for this area was then calculated from the relationship Regeneration Period : Rotation :: Area of Regeneration Block : Area of Whole Series by the formula,

$$\text{Regeneration Period} = \frac{\text{Area of Regeneration Block}}{\text{Area of Series}} \times \text{Rotation,}$$

instead of choosing the regeneration period and calculating the area of the block from the same relationship.

The period so fixed was then used for the calculation of an annual yield by Cotta' formula. This yield was, however, only prescribed for a fixed working plan period of twenty years, during which there was no expectation of regenerating the whole block. It was not intended to be applied for the calculated regeneration or exploitation period, nor for the regeneration period appropriate for the species and locality. At the end of the twenty years a new block was selected, including the reduced area of the partially regenerated compartments and the full area of such other compartments as were then ripe for exploitation, and a new yield was calculated.

The method is logical for a forest with a fairly normal age-class distribution, but has obvious dangers in other forests. It is still used in the forest of Compiègne, but has not been applied very generally.

B4. JUDEICH'S STAND MANAGEMENT METHOD[1]
This method was evolved by Judeich and Hufnagl in South Germany in the middle of the 19th century, and is very similar to the Single Periodic Block Method. It was devised for use in

[1] See page 192.

mixed hardwood and coniferous forests in hilly country which were divided into a number of Felling Series, each of which was exploited and regenerated by a shelterwood strip system of narrow strips in which seeding, secondary, and final fellings were made against the wind in a felling cycle. A suitable period was decided for regeneration and a list of stands was drawn up in order of their priority of need for regeneration.

The area of the Periodic Block was then calculated as

$$\text{Area of P.B.} = \frac{\text{Total Area of Felling Series}}{\text{Number of Periods in the Rotation}}$$

$$\times \frac{\text{Actual Average Age}}{\text{Normal Average Age } (\tfrac{1}{2} \text{ Rot.})}$$

The Actual Average Age was determined by multiplying the area of each individual stand in the series by its average age, summing the products, and dividing the sum by the total area of the series. Stands were then allocated to the Periodic Block from the list in order of priority of their need for regeneration until the calculated area of the block had been made up. Obviously, if there was an excess of old woods the actual average age of the series would be higher than the normal or half-rotation age, and consequently the periodic block could be larger than the normal block. If there was an excess of young woods the block calculated would be smaller than a normal block.

Example of calculation of size of block (Note that to save space the stands have been grouped into age-classes, but the principle is the same).

Age-class, years.	Area of Class, acres.	Average Age of Class × Area.	Average Age of Series.
1–20	400	10 × 400 = 4,000	
21–40	300	30 × 300 = 9,000	
41–60	700	50 × 700 = 35,000	
61–80	600	70 × 600 = 42,000	
	2,000	2,000)90,000	45 years

Regeneration Period = 20 years, hence number of Periods in Rotation of 80 years = 4. Area of Periodic Block = $\dfrac{2,000}{4}$

$\times \dfrac{45}{40} = 562$ acres instead of the normal area of 500 acres.

The final annual yield was, of course, the volume of the stands placed in the Periodic Block at the beginning of the period plus their increment for half the period, divided by the number of years in the period. In addition to this there would be an annual yield from thinnings required on silvicultural grounds outside the block.

CHAPTER

12

C. REGULATION BY VOLUME

The great majority of the methods devised for control of yield have sought to calculate the possible cut from measurement of the growing stock of a forest and its rate of increment. Such methods are applicable to irregular forests in which age-gradations or age-classes are not segregated in separate areas, but the individual trees of all ages are mixed together, as well as to regular forests. Some of the early volumetric methods were very theoretical and therefore dangerous.

C1. *Von Mantel's Method*[1] In the theoretical normal forest

illustrated in Fig. 1 G.S. $= I \times \dfrac{r}{2}$ when '*I*' represents the final

volume of the oldest or rotation age gradation, which is theoretically equal to the sum of the increments put on in a year by each age gradation, and therefore is the legitimate annual

yield. Thus, transposing A.Y. $= I = \dfrac{2 \text{ G.S.}}{r}$, which is known

as Von Mantel's formula. All that is required is the volume of the whole growing stock of the forest or felling series which can be obtained from sampling or from complete enumerations. It has already been shown that the actual growing stock which will be measured is usually less than the theoretical normal growing stock represented by the formula G.S. $= I$

$\times \dfrac{r}{2}$. Consequently the annual yield calculated from the for-

mula, which is merely the '*I*' of the original formula, will usually be less than the annual increment and therefore a conservative yield. It is a final yield, because the older stands measured which make up most of the volume have already

[1] See page 190 for Masson's similar formula.

been thinned. Therefore, when it is applied to an all-aged forest in which the annual fellings, carried out on a felling cycle, include final fellings and thinnings done on the same area at the same time, it will almost certainly be conservative, an advantage when management is first started in a forest about which so little is known that this method of yield regulation is the only one available. In such a case a conservative yield should be fixed for the first working plan period, during which more information is being collected. One great defect of the method is that it calculates the yield from the measured growing stock as if this existed in a normal series of age-classes, regardless of the actual age-class distribution. Thus, it will indicate the same yield from a forest entirely composed of immature stands as from one composed of overmature trees, as long as the measured volume of the forest is the same.

As long as these facts are recognized the calculation may still be a useful check on other methods of estimating the possible yield. According to the formula the increment per cent is proportionate to the rotation and can be calculated for any rotation by calling the growing stock a hundred, thus:

$$I = \frac{2 \text{ G.S.}}{r} = \frac{200}{100} = 2 \text{ per cent or } \frac{200}{80}$$

$= 2\frac{1}{2}$ per cent for rotations of 100 and 80 years respectively, which again might be used as a check on other methods of calculation.

C2. *Symthies' Modification* This modification is one of several devised to meet conditions when it is not practicable to measure the whole of a growing stock. In the tropics there is often a fairly high limit below which logs are not saleable, and limited staffs makes it important to simplify the work of enumeration as much as possible. In this country measurements are generally made to include all trees down to about 4 inches diameter at breast height ($3\frac{1}{4}$ inches quarter girth) and the volume of individual coniferous trees is measured right up the stem to the point at which the diameter falls to 3 inches. The volume of the stand not included in these measurements is negligible. When, however, nothing under, say, 12 inches quarter girth is saleable, it is sometimes the practice to enumerate only the trees of this size and above, and to measure the trees used for the construction of the volume tables, by which the tree count is converted into volume, only up to the points on their stem to which they contain saleable timber, say, 12 inches quarter girth again.

What is measured is not the growing stock, but a volume which can be called 'V', above a girth equivalent to an age 'x' which must be ascertained. An Indian forest officer called Symthies demonstrated that in such a case Von Mantel's formula A.Y. $= \dfrac{2\text{G.S.}}{r}$ could be modified to read A.Y. $= \dfrac{2V}{r-x}$ as can be seen in Fig. 6. In the figure the growing stock is represented by the triangle ABC, and the measured volume 'V' by the triangle XBD.

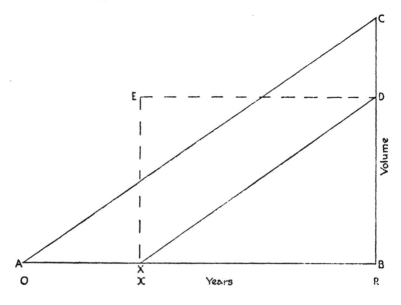

FIG. 6 SMYTHIE'S MODIFICATION OF VON MANTEL
ABC = Whole Growing Stock.
XBD = Volume Measured = V.
EXBD = Yield in Rotation — X Years = $2V$.

Annual Yield $= \dfrac{2V}{R-X}$

The yield in '$r-x$' years is equal to the rectangle EXBD,

which is equal to twice the triangle XBD, which is the measured volume 'V'.

Therefore the annual yield $= \dfrac{2V}{r-x}$.

This modification is even more dangerous than the original formula, which assumes the existence of a normal growing stock below the size of measurement. When the size of measurement is increased considerably this assumption becomes a very big one.

C (*b*). REGULATION BY COMPARISON OF ACTUAL AND NORMAL GROWING STOCKS

C3. *Austrian Method*[1] In 1788 an anonymous Austrian tax collector, who was concerned with the question of assessing woods for taxation, suggested a principle that became known as the Austrian Formula. He wished to find out what were the legitimate annual yields from forests containing various growing stocks so that they could be taxed. He suggested that the annual yield during a rotation ought to be the normal increment from a forest, assuming the forest to be in a normal state, plus or minus the difference between the actual growing stock of the forest and the normal growing stock from which the normal increment would be obtainable, divided by the rotation. That is to say, that less or more than a normal increment can be cut according to whether a forest is over- or under-capitalized.

This idea he embodied in a formula, or pair of formulæ, as follows:

If the actual growing stock is less than the normal,
$$\text{Annual Yield} = \text{Annual Increment, normal}$$
$$- \frac{\text{G.S., normal} - \text{G.S., actual}}{\text{rotation}}$$

If the actual growing stock is greater than the normal,
$$\text{Annual Yield} = \text{Annual Increment, Normal}$$
$$+ \frac{\text{G.S., actual} - \text{G.S., normal}}{\text{rotation}}$$

[1] See page 191.

For use in this formula the normal annual increment was taken to be the actual volume of the rotation-age age-gradation, the '*I*' of Von Mantel, and the normal growing stock was calculated from it as N.G.S. $= I \times \dfrac{R}{2}$. This, of course, suffered from many disabilities, such as that if the final age-gradation was excessive in relation to the rest both the normal increment and the normal growing stock would be over-assessed.

C4. *Heyer's Modification*[1] In 1841 one Karl Heyer, of Bavaria, thought it would be better to base the calculation on the actual increment rather than on the theoretical normal increment, and also that the state of growing stock should be brought to normal in a shorter period, '*a*' years, than that of the rotation, usually rather less than half the rotation.

In order to do this he estimated the increment of each stand in the forest, or at any rate of each age-class, for a short period of, say, ten years, and recalculated his yield at the end of each such period. At first he based the estimate on the mean annual increment of each class, obtained from measuring the volumes and dividing by age. Later Huber[2] substituted the C.A.I.s ascertained by increment borings. As all age-classes, including those in the thinning stages, are included in the calculations, the yield found is a total yield, not merely a final yield, as it was in the Austrian formula. Heyer's formula so modified is:

$$\text{A.Y.} = \frac{\text{Increment, actual in 10 years}}{10}$$
$$+ \frac{\text{G.S., actual} - \text{G.S., normal}}{a \text{ years}}$$

This second expression will be a minus quantity if the actual growing stock is less than the normal. The actual growing stock is, of course, obtained from enumerations, and the normal growing stock is now usually ascertained from yield tables (see page 25). In Switzerland, where the formula is still used as a check to other methods of calculating the yield, the

[1] See page 193.
[2] See page 212.

normal growing stock is calculated from the formula N.G.S. $= c \times R \times I$, in which 'c' is Flury's constant for the species, rotation, and locality (see page 22), and 'I' is the sum of the measured C.A.I.s of the various age-classes.

C (*c*). COMPARISON OF SUCCESSIVE ENUMERATIONS AND VOLUMES REMOVED

C5. *Biolley's Control Method*[1] This has also been called the method of continuous inventory, and the method of the measurement of the C.A.I. It is a modern practice which has been adopted for a number of European forests managed on the selection system, but is very intensive and has not been attempted in tropical forests, nor is it suitable for use with management involving clear felling of any considerable area. It is used in at least one privately owned beech forest in the Chilterns.

The details of the method vary considerably with local circumstances, but the essentials are:

(1) Measurement of all the trees in an area above a certain size at breast height, which is permanently marked, and the calculation of their volume ('V') by a local volume table.

(2) Measurement standing by the same volume table of the volume ('N') of all trees felled in the area until:

(3) Remeasurement of all trees on the area at the same marked point and of unmarked trees which have reached the size limit originally adopted, by the same volume table ('V2') after a fixed period of years, generally six to ten years.

Then Increment on the Area during the Period $= V2 + N - V$, and the annual increment is this amount divided by the period.

The increment can be calculated for a whole forest or for individual compartments. If the composition of the area is truly all-aged, with all size-classes present in the right proportions, then the increment of the area during a future period should be the same, provided the climate during the two periods is the same.

[1] See pages 213-220.

If, however, the forest is being converted from uniform system management to selection there will be large plots of even-aged trees which will become of measureable size at the same time, thus providing different amounts of recruitment in different periods. Also, in the second period there may be different proportions of trees in the different groups of size-classes with different rates of growth. In such a case it is necessary to keep the records of each group of size-classes separate and to calculate the increment for each group on the number of trees originally found in it, working downwards from the largest class, and to record recruitment to the smallest group separately from increment.[1] Then the increment percentages found for each group can be applied to the volume of trees found in the group in the second enumeration, and the total increment to be expected from the area in the second period, will be the sum of the increments for each group. Having decided the probable increment the forester then has to decide whether it can all be cut, whether some should be left to build up the growing stock, or whether some of the stock should be cut out in addition to the increment. He will also have to note which size-classes should be favoured and which reduced in volume.

Biolley did not do these things by reference to a theoretical normal growing stock and its size-class distribution. He cut trees on purely silvicultural grounds, believing that if in each compartment he could see adequate young regeneration, sufficient saplings and poles to allow selection of the best to grow on to larger sizes, there could not be much wrong with his growing stock. In selecting trees to cut he did not pay any attention to rotation, but left any healthy growing trees which were not interfering with others to put on value increment.

Under this system it is difficult to know what yield to prescribe for the first period, but as the period is short this does not matter greatly, and any of the volume methods can be used as a guide. The yield fixed is, of course, a total one as final fellings, thinnings, and tendings are all done together each year in the compartment or cutting section for that year.

Experience over sixty years in the Forest of Couvet, in Switzerland, where Biolley evolved his method, has shown

[1] See page 216.

that high yields are obtainable, of which increasing proportions are of large size, from a moderate volume of growing stock. He also found that the volume of the bigger size-classes which a selection forest can carry is larger in proportion to the smaller size-classes than in a uniform forest.[1]

[1] See pages 218 and 219.

CHAPTER

13

C (*d*). REGULATION BY SIZE CLASSES

This method was devised for uneven-aged and semi-uniform forests, and is based on the substitution of size-classes for age-classes for the purpose of yield regulation. When a regular forest can be divided into areas, each of which contains a definite age-class, it is possible to regulate final fellings during definite divisions of the rotation by area as in the Periodic Block Method (page 103). In an uneven-aged forest the age-classes are all mixed up and cannot be seperated by area. The size-classes roughly corresponding to periods of the rotation can, however, be recognized, and after an enumeration these size-classes can be grouped in a stand table which will show the separate volumes of the groups. Thus, in

C6. *The French Method of* 1883[1] (also called Mélard's Method, after the man who evolved it) three periods of the rotation were recognized. It was assumed that, in a rotation chosen so that in it the average tree will reach a certain exploitable size, during one-third of that rotation an average tree will reach approximately a third of the exploitable size, and during two-thirds of the rotation approximately two-thirds of that size. Suppose the average exploitable size is to be 60 cm. diameter, which corresponds with a rotation of 180 years, then all trees exceeding two-thirds of that diameter, i.e. over 40 cm. are grouped as Large Wood (*Bois Gros*), all of more than one-third, 20 cm., but less than 40 cm., as Medium Wood (*Bois Moyen*), and all below 20 cm. as Small Wood (*Bois Petit*).

The smallest group would not be measured, but the rest would be enumerated in, say, 5 cm. diameter classes as follows:

[1] See pages 195-199.

Diam. Class. cms.	Limits.	Group.
20	17·5–22·49	
25	22·5–27·49	Medium Wood
30	27·5–32·49	
35	32·5–37·49	
40	37·5–42·49	
45	42·5–47·49	
50	47·5–52·49	
55	52·5–57·49	Large Wood
60	57·5–62·49	
and above in 5 cm. classes if present		

Now as the volume of the Large Wood corresponds roughly to that of the trees of the last third of the rotation they can be cut and regenerated in a third of the rotation, just as if they were standing in a periodic block comprising one-third of the area of the forest. The average annual cut can therefore be the volume of the Large Wood divided by one-third of the rotation. But just as in the P.B. method the average tree will stand and grow during half of this period, so one-half of the annual increment of the Large Wood volume must be added to the annual yield. The Final Annual Yield can therefore be expressed as

$$= \frac{\text{Vol. of L.W.}}{\text{one-third of rotation}} + \tfrac{1}{2} \text{ annual inc. on Vol. of L.W.}$$

which corresponds exactly to Cotta' formula, in which Final Annual Yield

$$= \frac{\text{Vol. on P.B.}}{\text{Regen. Period}} + \tfrac{1}{2} \text{ annual inc. on Vol. on P.B.}$$

In the Periodic Block method thinnings were carried out in the blocks other than the one under regeneration, and the yield from these was additional to the final yield from the regeneration block. When this French Method of 1883 is used in conjunction with a selection system of silviculture, cutting is done on a felling cycle, and each year in the annual cutting section regeneration fellings of mature trees, thinning of pole groups, and tendings are carried out at the same time. There is therefore no separate thinning area from which to estimate the thinning yield, and it is necessary to calculate a total yield which can be harvested in the cutting section from all sizes of tree according to silvicultural requirements. Obviously, what should be thinned out each year is that part of the increment of the Medium Wood which is not needed to convert the present Medium Wood into the Large Wood of the next third of the rotation. This can only be estimated by careful study of the forest, its composition, stocking, and vigour. In a normal, well-stocked forest it should be about a third, but it may be less or even nil in some forests. The full formula for the total yield devised by Melard was therefore:

$$\text{A.Y.} = \frac{\text{Vol. of L.W.}}{\frac{1}{3} \text{ rotation}} + \tfrac{1}{2} \text{ annual inc. on L.W.} + \frac{1}{q} \text{ annual}$$

inc. on M.W. ($\frac{1}{q}$ being the fraction decided by inspection, $\frac{1}{3}$, $\frac{1}{4}$, or o).

There was, of course, no reason why a tree which is 60 cm. diameter at 180 years should have been 40 cm. at 120 years and 20 cm. at 60 years, but by measurements carried out over large areas of coniferous forest in the Vosges, French foresters found that the three groups thus formed did conform approximately to the three periods of the rotation. From the Vosges enumerations in 110 forests they accepted as normal stocking

L.W. 2,280 H. ft. p. acre
M.W. 1,360 ,, ,, = 3,640 H. ft. p. acre prop 5
,, 3

This they illustrated theoretically by reference to the growing stock triangle of the normal forest (see Fig. 7), which showed that the proportion which the volumes of the three groups should bear to each other was 1 : 3 : 5. The enumerations in the Vosges bore out that these proportions existed in what they considered satisfactory selection type forest. The growing stock triangle is, of course, based on the old fallacy of the final crop M.A.I. If, however, the proportions are based on a normal curve of production plotted according to the C.A.I.s for each age-class, that of the Large Wood to the Medium Wood is not very different, as can also be seen from Fig. 7. The Small Wood which is not measured and does not figure in the calculations at all is considerably smaller, but this does not matter.

When the method was first used and the enumeration of a felling series showed that the proportions of Large to Medium Wood varied considerably from this 5 : 3 ratio adjustments

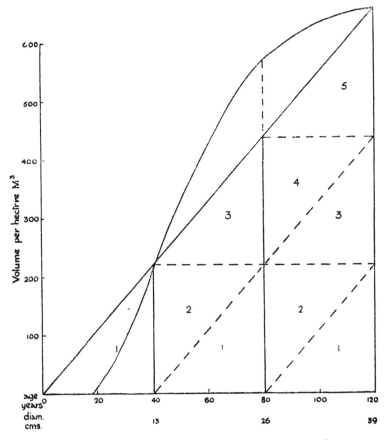

FIG. 7   DISTRIBUTION OF A NORMAL GROWING STOCK BY THIRDS OF A ROTATION.

		Size.	Classes.	Diameter, cm.	
		0–13	13–26	26–39	
Theoretical Stock calcu-	Vol. m³	4,373	13,120	21,867 =	39,360
lated by M.A.I. at Ro-	,, %	11·1	33·3	55·6	100
tation Age.	Prop.	1	3	5 =	9
Stock calculated from	Vol. m³	2,020	16,692	25,158 =	43,870
Yield Table.	,, %	4·60	38·05	57·35 =	100
	Prop.	0·42	3·42	5·16 =	9

(Based on E. Wiedemann's Yield Table for Spruce, Class II, Hanover, 1949.)

were made before the yield was calculated. Thus, if the L.W. was in excess and the M.W. deficient, one or more of the smallest diameter classes of the L.W. were transferred on paper to the M.W., and vice versa if the M.W. was deficient. Thus, if L.W. = 50,000 cu. ft. and M.W. = 30,000, the proportions were normal and the calculation for an 180-year rotation was:

$$\text{A.Y.} = \frac{50,000}{60} + \frac{50,000 \times 0\cdot01}{2} + \frac{1}{3} \times 30,000 \times 0\cdot03$$

$$833 \quad + \quad 250 \quad + \quad 300 \quad\quad = 1,383$$

(in normal conditions 'q' was generally taken as 3, the increment on the Large Wood as one per cent, and that of the Medium Wood as three per cent).

But if the enumeration showed L.W. = 62,000 and M.W. 18,000, there was an excess of 12,000 cu. ft. of L.W. and a deficit of 12,000 of M.W. Therefore the enumeration stand table would be examined to see how far this could be overcome on paper. Suppose that the 40 cm. and 45 cm. diameter classes contained between them 8,000 cu. ft., these would be transferred from the L.W. group to the M.W. group and the calculation became:

$$\text{A.Y.} = \frac{54,000}{60} + \frac{54,000 \times 0\cdot01}{2} + \frac{1}{3} \times 26,000 \times 0\cdot03$$

$$900 \quad + \quad 270 \quad + \quad 260 \quad\quad = 1,460$$

a larger annual cut which will reduce the proportion of the L.W. from which most of it will be realized, but not so rapidly as if no transfer had been made. The calling of the 40 cm. and 45 cm. diameter classes Medium Wood for the purpose of the calculation will not affect the marking of trees to be cut which would be done purely on silvicultural grounds up to the required 1,460 cu. ft. a year. The effect of the transfer is merely to cause a lower yield to be calculated than otherwise would be, and to allow for very little thinning out of the deficient Medium Wood to be done.

More recently this method of transfer has been given up, and flexibility in suiting the prescribed cut to the size group distribution is obtained by adjusting the variables, the two increment rates and 'q'. Instead of using the more or less standard rates of 1 per cent and 3 per cent, the actual rates for the forest are calculated, and these will, of course, reflect the density of the stocking, if growth conditions remain constant.

It is of interest to note that when the method was introduced in 1883, no increment, either of the Large Wood or of the Medium Wood group, was allocated for removal. The result was that as much cutting as was silviculturally desirable could not be carried out within the limits of the prescribed yield, regeneration was retarded, and the rotation was prolonged.

C7. *The Regeneration Area Method*[1] (*Méthode à Quartier de Régénération, or Mélard's Method of* 1894) The same method of yield regulation by volume of size-groups was later applied to forests worked not on selection silviculture, but on what may be termed a quasi-uniform system. These forests were generally coniferous mountain ones, particularly in the Jura, in which, after treatment for some years by the selection system, attempts had been made to enforce the uniform system by periodic blocks (ordered in 1853). Because of their irregularity and because of storms which blew down trees and started regeneration in scattered compartments of different periodic blocks, little progress was made towards uniformity. Also, because only the final yield was controlled, there was no certainty that when the volume of mature trees felled in the thinning blocks on purely silvicultural grounds or blown down was added to the final yield, the total would conform to the real annual possibility of the forest.

It was therefore decided that it would be better to calculate an annual total possibility, based on a complete enumeration of the whole forest down to a diameter of one-third of the average exploitable diameter, by the method devised in 1883. The probable mean annual increment of the Large Wood group was included and a portion of the increment of the Medium Wood group, when enough was known about the forest to decide what the proportion ought to be.

The total yield having been fixed the next step was to organize the progress of as uniform regeneration as possible. For this purpose all compartments, or sub-compartments, were allocated to two groups for a working plan period which should not exceed twenty years in length. In the first or Regeneration Group were placed those compartments whose regeneration ought to be carried on or started during the working plan period, that is to say those whose regeneration was urgent, in order of urgency. The remaining compartments were placed in a Thinning Group. The Regeneration

[1] See page 199.

Group or Area was not a periodic block which was due to be felled and regenerated in a definite regeneration period, but a collection of compartments in which regeneration operations would be concentrated during the current working plan period. The Thinning Group or Area was the collection of compartments in which thinning and tending operations were to be carried on on a thinning cycle.

The total yield calculated by Mélard's formula can be divided between the regeneration fellings in the Regeneration Area and the thinnings in the Thinning Area in two different ways:

(1) The method specified in 1894 was to prescribe by area, on a chosen thinning cycle, the annual thinning coupes, and to thin these each year in accordance with the silvicultural rules prescribed before any other cutting was done. The volume of these thinnings and of all windfalls, counting only trees of the minimum size enumerated and over, was ascertained and subtracted from the prescribed total yield. The balance was then felled in regeneration fellings in the Regeneration Area. This method sometimes did not allow enough regeneration felling to be done to bring on the young trees, particularly when there had been heavy windfalls in the Thinning Area.

(2) To overcome this difficulty a later method was to allocate a definite portion of the total yield to each group, and in each group to cut the allocated volume each year in a prescribed order of compartments, counting any windfalls in each group as part of the allocated yield of that group. This meant abandoning any prescription of area to be thinned each year, and only thinning as large an area as the allocated volume would last for. This has sometimes meant that the thinning cycle has got badly into arrears.

The plan is drawn up for a definite short period of, say, ten years. However, under the second method of allocating a definite volume of the total yield to regeneration fellings, the duration was sometimes estimated as follows. The volume of exploitable trees in the Regeneration Area was ascertained from the enumeration figures and divided by the annual allocation for regeneration fellings. The estimated increment which should be put on during the number of years so ascertained was then found and sufficient years added to the period to cover its felling, thus:

Annual Allocation to regeneration fellings = 1,000 H.ft.
Volume of exploitable timber in Regeneration Area = 15,000 „

There are therefore 15 years' cutting, for half of which increment will accrue, i.e. $15,000 \times 0 \cdot 01 \times \dfrac{15}{2} = 1,125$ H. ft., or another year's cutting. The yield determination can therefore continue for 16 years, after which the plan will need revision.

This calculation of a period is condemned by modern French foresters as destructive of the matching of the rate of regeneration fellings with the progress of regeneration, which is only obtainable from re-enumerations and recalculations of the felling rate at short intervals. They say that it almost converts the Regeneration Area Method into the Single Periodic Block Method; that compartments containing the bulk of the large timber should always be placed in the regeneration area, and remain there till one by one they are passed out fully regenerated to the thinning area, some probably after several working plan periods.

Each time the plan is revised the increment rate of the growing stock is calculated and compared with the cut, thus:

Growing Stock enumerated 1940, after annual felling = 85,000 H. ft.
Total Cut in the 10 years 1931–40 = 15,500 „

—————

100,500 „
Less Growing Stock enumerated 1930 = 80,000 „

—————

Total Growth in 10 years = 20,500 „
Mean Growing Stock = $\dfrac{80,000 + 85,000}{2}$ = 82,500 „

Increment Percentage per annum = $\dfrac{20,500 \times 100}{82,500 \times 10}$ = 2·48 per cent

If it is considered that the forest is likely to continue growing at this rate then the annual increment of 2·48 per cent on the 1940 growing stock of 85,000 H. ft. would be approximately

2,100 H. ft., and this figure would be used as a guide and a check to the calculation of the annual yield by Mélard's formula for the next period. The variable 'q' would probably be adjusted so as to bring the solution of the formula to about 2,000 H. ft.

There is, however, a danger in adopting this figure of apparent production for a series composed of uniform or semi-uniform stands. In a truly all-aged forest the volume calculated from a volume table from the diameters or girths of the trees should bear practically the same relationship to the true volume at all times. In the *Méthode du Controle* the use of the same volume table for all calculations is stressed for this reason (see p. 215). In uniform or semi-uniform forests the average age of a series is unlikely to be the same at each enumeration, and the form of various stands may have been affected by the intensities of the thinnings applied to them. Consequently the height/diameter relationships and the form factors of the trees are likely to be different at each enumeration, and therefore the ratio which the volume table volume bears to the true volume will vary. Any considerable difference in apparent production between one period and another should, therefore, be viewed with suspicion.

In this method the compartments of the Regeneration Area were coloured blue on the management map and sometimes were referred to as *Le Quartier Bleu* (The Blue Area). The name, however, has also been applied loosely to the current regeneration block under a periodic block method, when it is really an *Affectation Bleu*, and so should be avoided as likely to cause confusion. Sometimes the Thinning Area is divided into two, *Le Quartier Jaune* (The Yellow Area) containing fairly mature stands, and *Le Quartier Blanc* (The White Area) containing the youngest stands, for which different thinning regimes are prescribed.

C8. *Yield Regulation by Number of Trees, Brandis' Method* Regulation of cutting by size-classes on a rather different basis, that of the numbers of trees which can be removed, was evolved to meet the need for exploitation of large areas of tropical forest. Practically nothing was known about these forests, and it was necessary to allow some production from them with a minimum of damage to them, while information was collected which would make more intensive management possible.

In 1856 Brandis had to regulate cutting of teak in Burmese forests over which concessionaires had been granted cutting rights. Only teak, which represented about ten per cent of the growing stock, was saleable. As in many untouched tropical forests there was an excess of large trees of the valuable species over the medium sized and small ones. If all the merchantable trees were to be cut out as fast as the concessionaires wanted it would have been a long time before the forests would be productive again, even assuming that regeneration of the valuable species appeared, could establish itself and compete successfully with all stages of the growth of the useless species.

Brandis decided that only teak trees over six feet in girth ought to be cut and proceeded to carry out enumerations by strips to determine at what rate they could be cut. He enumerated the teak in three girth classes, Class I, over 6 feet; Class II, $4\frac{1}{2}$–6 feet; and Class III, 3–$4\frac{1}{2}$ feet. Later larger numbers of girth classes were differentiated and the exploitable size was varied according to localities. Brandis was fortunate in that teak have annual rings so that by counting these he was able to determine average figures for the number of years a tree took to pass through each class. He also estimated the approximate percentages of trees in each class which were likely to be available for cutting at six feet girth or over, making allowance for those which would die or be suppressed, or which would have to be left standing as seed-bearers in places where no immature teak would be available to leave as seed trees.

If he found that there were in a forest, say, 5,000 Class II trees of which eighty per cent, or 4,000 trees, could be expected to be available for cutting eventually as Class I trees, and had ascertained that it took forty years for the average teak tree to pass through Class II, then obviously 4,000 harvestable trees would reach cuttable size in a period of forty years at an average rate of 100 trees a year. In other words, the first thing he did was to calculate the average annual recruitment to Class I by dividing the effective number of Class II trees by the number of years it would take them to reach Class I. In a forest in which size-class distribution was normal the cutting of this number of trees a year would maintain normality but, of course, these forests were not normal, so the position had to be examined further. For reasons of economic working felling cannot be carried out all over a large forest every year, and the only practical method is to concentrate felling each year in an annual cutting section which is visited on a convenient cutting cycle. This means that there must be a working stock of trees of exploitable size on the ground to enable the annual recruitment to exploitable size of the whole forest to be cut from a portion of the area each year.

Suppose for the sake of argument that felling of the full number of the recruitment rate is to be done over the whole forest each year, i.e. that the felling cycle is one year. Then if cutting is done before the growing season none of that year's recruits can be cut, whereas if felling is done after the end of the growing season all of them can be cut. On the average when felling is done throughout the year half the year's recruits can be cut, and a number of exploitable trees equal to half the annual recruitment rate will be needed as a working stock at the beginning of the year. These will be cut and replaced by the uncut half of the year's recruitment to form the working stock for the beginning of the next year. Similarly, with a felling cycle of two years only half the annual recruitment of half of the forest (the first annual cutting section) can be harvested in the first year, but one and a half times the annual recruitment on the second half will be ready for cutting in the second year. Thus the full recruitment rate for one year over the whole forest could be cut in recruits during the two-year felling cycle period, and the balance, also one year's full recruitment would have to be present at the beginning of the period (i.e. annual recruitment rate × $\frac{1}{2}$ the felling cycle of two years).

Fig. 8 shows how an annual recruitment rate of 100 trees can be harvested on a ten-year felling cycle with the help of a working stock of 100 × $\frac{10}{2}$ = 500 trees of exploitable size. It also shows how the uncut half of the recruitment during the

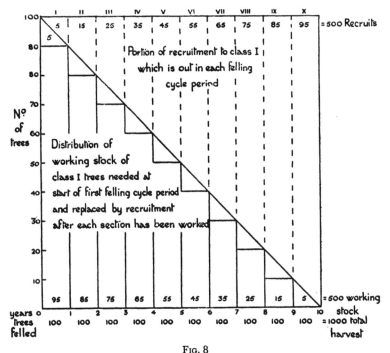

FIG. 8

Diagrammatic representation of the harvesting of annual recruitment of 100 trees a year on a ten-year felling cycle from equiproductive annual cutting sections.

$$\text{Working Stock} = \text{Annual Recruitment} \times \tfrac{1}{2} \text{ Felling Cycle}$$
$$= 100 \times \frac{10}{2} = 500 \text{ trees (Class I)}$$

($+ \tfrac{1}{2}$ Annual Recruitment if all cutting is after end of growing season, or $- \tfrac{1}{2}$ A.R. if all cutting is before start of growing season, i.e. $\pm$ 50).

cycle is built up in the cutting sections after they have been exploited to form the working stock of 500 trees needed in the next cycle. Thus, when the annual recruitment rate to exploitable size and the length of the felling cycle are known it is possible to calculate the number of trees now exploitable, i.e. Class I, which must be left on the ground to enable the full recruitment rate to be exploited. Any surplus of Class I trees can be liquidated over any convenient number of years. If the exploitable trees are likely to deteriorate quickly the surplus can be cut out during the first felling cycle. If doing so would mean a serious drop in yield during the second cycle

131

the possibility of spreading the liquidation of the surplus over two or more cycles should be considered. It must be remembered that it will be possible to cut out the oldest trees and to store new recruits to Class I in their place.

Suppose that the following data had been collected:

Class.	Girth, ft.	Gross no.	Years in class.	Percentage likely to be cut as Class I.	Net no. of harvestable trees.
I.	over 7	31,580	—	95	30,000
II.	6 –7	21,175	26	85	18,000
III.	4½–6	38,570	37	70	27,000
IV.	3 –4½	64,000	32	50	32,000
V.	1½–3	112,000	30	25	28,000

Suppose also that it is believed that it takes 25 years for a tree to reach 1½ feet in girth, making a provisional rotation of 150 years for the average tree of exploitable size of over 7 feet. Also that a felling cycle of 30 years is to be adopted, dividing this rotation into five felling-cycle periods.

Recruitment to Class I during first Felling-cycle Period During the first 26 years all the 18,000 Class II trees would reach exploitable size. During each of the next four years the average number of Class III trees which have been reaching Class II each year will pass into Class I.

Thus, Total Recruitment to Class I in 30 years

$$= 18,000 + 4 \times \frac{27,000}{37}$$
$$= 20,920$$

Average Annual Recruitment Rate
$$= 697 \text{ or, say, } 690$$

To harvest 690 trees a year on a 30-year felling-cycle a working stock of $690 \times \frac{30}{2}$ Class I trees $= 10,350$ are needed on the ground.

There are available 30,000 Class I trees, so the surplus $30,000 - 10,350 = 19,650$ can be felled. If this is done during the first felling-cycle period then $\frac{19,650}{30} = 655$ trees a year can be cut in addition to the recruitment rate of 690, making an annual yield of 1,345 trees. If the surplus is spread over two

felling-cycle periods the annual yield will be $690 + \dfrac{19,650}{60}$
$= 1,017$. Consideration must now be given to what the yield is likely to be later in the rotation and whether, if the working stock of Class I trees is reduced to 10,350 trees, future annual recruitment rates will be able to be harvested.

It is possible to calculate the average recruitment rate for the whole period during which the smallest enumerated tree will reach Class I. This period is 125 years and there are now 105,000 trees below Class I, so $\dfrac{105,000}{125} = 840$ will be the average annual recruitment rate to Class I during the next 125 years.

To harvest this a working stock of $840 \times \dfrac{30}{2} = 12,600$ trees will be needed in Class I all the time, so the surplus is only $30,000 - 12,600 = 17,400$.

To decide how long ought to be taken in liquidating this surplus the total number of trees which can be felled in the 125 years, leaving the working stock of 12,600 at the end for future use, can be found. The annual rate will be $\dfrac{135,000 - 12,600}{125}$
$= 970$. The first felling-cycle period can now be reconsidered,

Average annual recruitment rate during first cycle	= 690
Surplus of Class I working stock over the average stock required = $30,000 - 12,600 = 17,400$, if spread over two felling-cycle periods	
annual cut of surplus $= \dfrac{17,400}{60}$	= 290
Total Annual cut	= 980

which looks about right, so that a cut of 950 or 900 trees a year could be prescribed, depending on how much reliance is placed on the figures.

It is wise to be conservative in fixing a yield for a forest about which little is known, but it must not be forgotten that under-cutting is liable to hamper regeneration and growth rates generally. Re-estimations of the stock at as frequent intervals as can be afforded will show what is happening and allow new estimates of the permissible yield to be made.

In order to enable a simple rule, which will facilitate the marking of the prescribed number of trees and the leaving of the desired number of the surplus, to be included in the cutting rules, it is sometimes a convenience to express the yield as a percentage of the exploitable trees which will be found in the forest during the cutting cycle. The number of such trees will be the number of Class I trees at the beginning of the cycle plus half the recruitment to Class I during the cycle (i.e. that part of the recruitment which will be available for cutting).

Thus, in the example just used there were 30,000 Class I trees at the start of the first cycle, and 20,920 other trees would become Class I during the 30 years. Hence the total number of trees over 7 feet which will be found during the course of marking operations carried out on the various cutting sections will be $30,000 + \dfrac{20,920}{2} = 40,460$. If the average number of trees to be cut each year had been prescribed as 900, or 27,000 for the 30 years, these could be expressed as a percentage of the 40,460, i.e. $\dfrac{27,000 \times 100}{40,460} = 66$ per cent. The felling rules could then include one to the effect that of each three trees of exploitable girth found two could be marked for felling, and one must be left, subject to the silvicultural felling rules.

In practice, of course, the distribution of exploitable trees will not be even, and it will be difficult to divide the forest into equi-productive cutting sections. If the areas of the annual cutting sections are fixed, the number of trees which can be felled will vary each year, though the total prescribed for the period should be available, unless the enumerations or calculations are wrong. Alternatively, it is possible to harvest the prescribed number of trees each year, working in strips starting from where the previous year's felling ended over whatever area is needed to obtain them. In this case if the enumeration or calculation were wrong the whole excess or deficit of trees would occur in the last year of the cycle.

The calculation of the percentage of exploitable trees which may be cut in common in India, particularly for Sal (*Shorea robusta*) selection forest, where it is often done by what is called the Safeguarding Formula. This is generally the recruitment rate to Class I for the cycle expressed as a percentage of the available Class I trees, and rounded off to a convenient figure. No account appears to be taken of any existing surplus in Class I, though an arbitrary modification can be made.

The Brandis method may well be of value for yield regulation in some colonial tropical forests in which size-class distribution indicates that selection silviculture is likely to be successful. In spite of the lack of annual rings in many tropical species their rates of growth are now being determined.

Where volume tables exist it is possible to convert a yield calculated in numbers of trees into volume by estimating the average size at felling of the prescribed trees. In a forest with a normal working stock of Class I trees this would be the minimum exploitable girth plus the girth increment estimated for half a felling-cycle period. With a large stock of over-mature trees it might be necessary to make an allowance for decadence in the trees which were over the exploitable girth at the beginning of the cycle. In the absence of volume tables the numbers of trees to be cut have sometimes been converted into square feet of basal area. In the 1929 plan for the Mu Division of Burma the average basal area of all trees cut during the previous ten years was used for this calculation. The advantage of expressing yield in volume, or in basal area, is that marking can be done on silvicultural grounds, leaving large, vigorous trees to put on value increment, and removing unsatisfactory smaller trees as part of the yield. However, the adjustment of canopy in a mixed tropical forest depends more on the cutting or poisoning of the numerous unmarketable species and of climbers than on the removal of the few merchantable species. Selection among the latter cannot go much beyond seeing that healthy seed-bearers are left adequately distributed.

C (*e*). YIELD REGULATION BY DEVELOPMENT, TENDING, OR CUTTING CLASSES

These methods are common in Scandinavia. Sometimes in Norway all the stands of a forest are classified in classes which represent roughly defined age-classes. The proportions which each class bears by area to the proportion it should occupy in a normal forest are studied, and the felling rate for each class is so fixed as to bring its proportion by area to normal.

C9. *Swedish Cutting Class Method* In this, stands are assessed by eye for allotment to so-called Tending Classes, as follows:

Tending Class I. Cleared ground and ground incompletely regenerated.
  „   „   II. Young stands in the cleaning and wolfing stages.
  „   „ III. Older stands in thinning stages.
  „   „   IV. Still older stands putting on good value increment.
  „   „   V. Stands of any age not making the best use of their sites, or whose increment is falling off, and so requiring removal and replacement.

Classes III, IV, and V are divided into age-classes, their volume per acre and total volume are listed, and an estimate is made of the felling per cent, which should be applied to each age-class for the next ten years. By a felling per cent is meant a percentage of the present volume of the growing stock in the class; for instance, three per cent of the present stock to be cut in thinnings in Class III.

Tables are used to convert this felling per cent calculated on the growing stock at the beginning of a period into an exploitation per cent, which takes into account both the annual increment and the annual felling. This exploitation per cent is the annual out-turn, calculated after the year's growth has finished, as a percentage of the growing stock at that time, before the winter felling starts, for the appropriate felling per cent. The relationship between the exploitation per cent and the felling per cent is calculated on the assumption that the timber capital in both cases will be the same after ten years. The local officer is left to harvest this volume where and how he considers wise on silvicultural grounds each year. The silvicultural system practised most commonly in Sweden is one of group-selection, and the working plan period is almost always ten years.

CHAPTER

14

It is obvious that the method selected for regulation of the yield in any particular area will depend on the type and form of forest, the intensity of working proposed, the resources and time which can be devoted to the enumeration and increment studies, the availability of research data, such as yield tables, and to some extent the system of silviculture to be practised.

For intensive working in which all increment put on by the growing stock will be harvested, careful measurement of the present volume of each age- or size-class and its rate of growth will be worth while. Intermediate yields and the types of produce which will be available from them will be quite as important as the final yields. In previously unmanaged forests containing a preponderance of unmarketable species and of over-mature trees of the marketable species, the aim of management is generally to spread out the harvesting of the merchantable timber until part of a new and more profitable crop which is to be established reaches exploitable size. This new crop may be obtained from natural regeneration tended so as to favour the valuable species, with or without supplementary enrichment planting, or from artificial regeneration in compensatory plantations established on part of the area exploited. The time over which the removal of the old crop has to be spread cannot be much less than the intended rotation of the new crop, and where most of the original trees are mature it will be difficult to avoid loss by decay, unsalvageable windfalls, etc., and seldom possible to harvest much increment. To do so would necessitate working over the whole forest on a short felling-cycle, cutting out first only the most over-mature trees and gradually reducing the permissible girth limit for cutting. Combing large, roadless tracts for possibly one tree per fifty acres would not be economic, and a long felling-cycle with a low minimum exploitable girth often has to be adopted.

As already explained, Nigerian foresters apply area control to this type of forest, laying out annual coupes of equal size, fixed by dividing the area of a felling series by the assumed rotation of the future crop. Advance regeneration is established in these coupes, useless trees are cut or poisoned, and then all merchantable trees are felled to leave a uniform young crop. There is no assurance of equal annual yields from the coupes in which the distribution of valuable species may vary considerably.

In Kenya practically unrestricted selective felling by saw-millers has been allowed over the natural forests and compensatory plantations, mainly of exotic softwoods, are being established on something like a quarter of the area on the best and most accessible sites. These are expected to be mature in about forty years from planting, but no estimate of the original stock of marketable timber was made and the rate of planting was governed by the availability of staff and funds at various periods, with the result that a very abnormal distribution of age-classes exists in the new plantations. This will have to be corrected by starting to fell the oldest crops well before maturity, a procedure which will probably be necessary in any case in order to bridge a gap between the exhaustion of the original forests and the maturing of the oldest plantations.

In the Federation of Malaya, where a rotation of seventy years has been adopted for Dipterocarp forest worked on a uniform system, the accessible Dipterocarp area of each State is treated as one working circle. Each State Forest Officer draws up annually a felling plan of scattered coupes of various sizes, suitable for the contractors and saw-millers interested in buying timber. These amount to one-seventieth of the total area of the circle. Where the timber firms are increasing in size and stability, separate felling series, each of which will provide seventy annual coupes, are being formed for allocation to specific concerns and their successors in tittle, to assure them steady supplies of raw material from one locality.

In Uganda a maximum permissible annual volume yield is fixed for each felling series, but it refers only to those timbers for which there is an assured market and which have attracted the sawmiller to apply for a felling licence. The volume of the growing stock in a series is estimated from sample enumerations and from records of output per acre from any part of the same or of a similar forest which has been exploited. A provisional rotation at the end of which it is expected that the new crop to be established will be merchantable is fixed, and is divided into two felling cycles which are decided by the size-class distribution of the existing valuable species.[1] For instance, a provisional rotation of 120 years might be divided into two cycles of 80 and 40 years respectively. During the first 80 years the annual yield of the assured market trees will be the total volume of these species over a suitably selected girth limit divided by 80. This yield is harvested by working in contiguous strips from a prescribed starting line, felling all trees of the prescribed species over the prescribed girth on whatever area is necessary to provide it. In the annual coupe so determined the licensee is encouraged to fell as many trees of the unrationed species as the forest officer is willing to mark on silvicultural grounds. The girth limit for the rationed species is fixed with the intention that during the second felling-cycle the trees below the limit left standing in the first cycle will provide another forty years' cutting of approximately the same annual volume from coupes of double the size. Regeneration is either by supplementing and tending natural regeneration in the annual coupes exploited to obtain the yield, or by compensatory planting of part of the forest. The maximum annual volume yield fixed in the original plans is amended at revisions in the light of the actual output from the area worked over to date. As the plans are revised every ten years the fact that the original regulation of the yield was based on inadequate data is not very serious, and does not lead to dangerous over-cutting of the felling series area as a whole. Excess cutting had to be allowed during the war years 1940–46 and is being covered by reductions spread over the rest of the provisional rotation. This is merely a modification of the area control method, and area control in one form or another is practised in almost all of the previously unmanaged forests of the tropical colonies, just as it was in the forests of France when management was first introduced there. If the regenera-

[1] See Appendix II for an elaboration of this system.

139

tion which follows it is successful, age-class series are built up which will be the basis for more scientific forestry in the future as they have been in France.

The results of the enumeration surveys, etc., carried out will have been summarized already in Part I, Chapter IV, and in Part II of the plan the working plans officer should show briefly how he has used them to determine the rate of felling and regeneration. He should prescribe the output in terms of volume or units, or in terms of area for a convenient period, and state the average annual yield at which working should aim. When only a final yield is fixed an estimate of the intermediate yields which will be removed in thinnings from the areas other than those to be regenerated in the period should be given. This may be based on previous records, dummy markings on sample area, or on yield tables. A complete forecast of the annual production of all merchantable forms of forest produce is required. How this has been drawn up should be explained clearly and concisely, so that any one reading the plan can follow the calculations made and judge their probable reliability.

CHAPTER

15

This section contains the detailed instructions about how and where the yield is to be harvested. It is one of the most vital parts of the plan. Tables should be prepared showing for each type of cutting, e.g. regeneration felling, improvement felling, or thinning, which is to be carried out, the location[1] and amount of the work to be done during each individual year of the plan. A separate table may be drawn up for each Felling Series with columns for the types of cutting, or separate tables may be drawn up for each type of cutting, with columns for each felling series. The layout of the tables may vary according to the system of management, but the first column will normally be for the year, the next for the compartment number, then the area to be worked over if the yield is regulated by area, or the volume to be cut if the yield is regulated by volume.

The tables are also of value in helping the working plans officer to estimate the revenue to be expected and the expenditure likely to be incurred in cutting operations each year. Bourne suggested that the value and costs should be included in the tables, but there is a separate chapter for finance later on and it may not always be wise to risk complicating what should be simple tables of prescribed cuttings. The same applies to the addition of control columns which Bourne suggested. The possibility will be considered later, when the method of control is discussed. The sort of table it is possible to draw up for all the special plans which are dealt with in this chapter of the working plan could have four functions:

(1) Simple prescription of operations to be carried out.

(2) Forecast of results of operations, i.e. volumes likely to be obtained from areas prescribed under area control, or of

[1] But see page 142.

141

area likely to be worked over to obtain a prescribed volume, and of estimated expenditure and revenue on and from the operation.

(3) Records of actual results obtained by each operation, volumetric and financial.

(4) Comparison of the amount of work prescribed with that actually done, analysed to show whether the actual operation of the plan is progressing as arranged, and if not what arrears or excess of felling, planting, cleaning, etc., there are at the end of any year. This is known as control of the plan.

In the present section the tables have only one function, to tell the executive officer just what fellings he is to make each year, where, and how much he is to cut. Probably a simple prescriptive table in the text and a copy or copies with additional columns for forecasts, records, and controls in the appendix will often be the best solution.

There are, of course, cases in which it is not the policy to fix definitely the actual times and locations of all harvesting and silvicultural operations in advance, even for the comparatively few years of a modern working plan period. For instance, when use is to be made of advance regeneration likely to occur after good mast years, heavy thinnings may be prescribed for as many stands in the pole stage as possible during the mast year and the one following, or it may be prescribed that all regeneration which appears in gaps opened by wind, etc., shall be tended. In such cases it is important that machinery be provided for recording just what work is started where, and for bringing the areas treated into the schedules of prescribed tending operations to ensure that proper attention is given to them at the right times in the future. If this is not done groups of regeneration which have been partially opened up or sub-compartments in which seedlings have been obtained in accordance with the intentions of the plan may be overlooked.

It is often wise to leave a great deal of discretion to the officer in charge of a forest, but it is very unwise for him to be allowed to carry the details of his management in his head only. In a small forest in which he has spent many years a skilled forester may be able to do this but, even so, if he has to leave suddenly and a new officer arrives at the beginning of a busy season, the latter will have difficulty in finding all the sites in which work needs doing in time to avoid losing a year. In a large forest district, such as is normal in the tropics, where transfers and sudden illnesses are all too likely, everything must be put on paper as soon as it is started and its continuation prescribed, so that an officer taking over in a hurry can find in the schedules of operations just what work has to be done. He may have a dozen plans to execute and no time in which to study them, but if the schedules have been kept up to date he can give the necessary instructions for the continuation of the work of each without loss of a season.

Sometimes a working plan will prescribe that sites for certain operations shall be selected in certain compartments each year, for instance, that one or more groups amounting to a certain area shall be planted up in the more open parts. The general principles on which the locations shall be selected will have been laid down in the plan, but in cases such as this it ought to be a rule that the working plan officer shall actually mark out on the ground the sites for the first year. This will provide a practical example of the way in which he intends the instructions to be applied.

Similarly, if a prescription for harvesting a yield, either final or intermediate, is not an entirely straightforward one, such as clear-felling an area or cutting a volume from an unrestricted area on certain clearly defined principles, the working plan officer should mark the first year's yield himself. This will not only demonstrate his intentions, but will be a check on his own prescription. It may well be that in attempting to carry it out he will find that it will not work, for instance, that the volume prescribed to be taken from a certain area cannot be obtained by marking in accordance with the cutting rules he has laid down. In such a case he will have to modify his prescription and will be saved from having signed an unworkable plan. The suggestion that no officer should draw up a plan which he is not going to have to operate has already been mentioned. The adoption of this is not possible, but certainly a working plan officer should do all he can to assure himself and others that his plan can be operated.

When skilled staffs of adequate size are available for forests of which the soils, ecology, and silviculture have long been the subject of close study, intensive and complicated management can be carried out. When, however, one officer with a few semi-skilled subordinates has to try to introduce some order into a large tract of little known jungle, it is wise to plan simple and straightforward working. By scientific standards the results may be crude, but refinements can be introduced later. If you have neither a workshop nor mechanics but want to move some soil it is better to build a wheelbarrow than to try to construct a lorry.

This section must include Cutting Rules for final fellings, thinnings, and other types of cutting, either as detailed prescriptions or as guiding principles for selection, marking, felling, etc., of the trees from which the yield is to be obtained.

PLAN OF CULTURAL OPERATIONS

Bourne called this the Plan of Formation, which seems rather too narrow a term. In this section must be recorded prescriptions of all operations needed to renew, replace, and tend all the forest crops up to the time when remunerative cutting prescribed in the previous section becomes possible.

Here again, a table or a series of tables should be devised which will tell the executive officer just what he has to do where each year, including the tending of operations started in locations not prescribed in advance. For instance, in a working circle to be treated by clear felling and artificial regeneration, the cleared annual coupe to be planted each year will be tabulated against the year, and further columns provided for the coupes which will need weeding and beating, cleaning, wolfing, etc., in the same year. In cases when pre-exploitation treatment is needed, as under the Nigerian Tropical Shelterwood System, the table would have to start for each coupe some years before the main fellings in it were due, by prescribing the year in which the first operation intended to encourage natural regeneration had to be carried out.

All cultural operations not directly remunerative must be provided for. If plants will be needed for artificial regeneration, their source and method of procurement must be prescribed. If fencing against deer, rabbits, etc., local drainage operations, local fire-protection, etc., are needed they must be prescribed.

The remarks already made about estimated costs, actual costs, and control columns apply also to the tables of this section.

Certain expenditure, such as that on the replacement of the cut, is inevitable, and the adoption of the cheapest means will not always be the most economical in the end. French foresters are often willing to wait for a number of years to obtain natural regeneration, but Scandinavian foresters say that they cannot afford to wait and so lose production from valuable soil. They generally plant up if natural seedlings do not come in in sufficient quantities at once. Sometimes they will even decide that the natural plants are not of a sufficiently good strain, and will plant seedlings raised from selected trees through what appears to be adequate natural regeneration.

In the tropics it is particularly important to secure canopy formation to control weeds as early as possible, and complete hoeing of a planted area several times may actually save money in the end, besides producing faster-growing and better stocked crops than weed slashing or spot weeding.

There are, however, many things on which a working plans officer would like to spend money, and he will have to consider the whole financial position very carefully before selecting those which he will prescribe. Forestry should be a paying proposition, and there will be few private owners who will not expect to make money out of their woods. States may be willing for a while to invest money in their national forests, but will expect it to be used economically and wisely. The forester who can show a profit on his activities will always be in a stronger position than any other to obtain funds for investment in his forests.

When the working plans officer is dealing with his cultural prescriptions he will find the notes he made in pencil in his Compartment Descriptions of the things he thought needed doing in each to be of great assistance. He will select from them those which are essential during the working plan period, and replace his notes by definite prescriptions. In regard to cultural operations generally, excessive rigidity of prescription should be avoided. For instance, a thinning-cycle of five years may have been selected as the best average interval between successive thinnings, and the thinning table may show that

Compt. 8 will be thinned in 1950 and Compt. 9 in 1951. When 1950 comes, Compt. 9 may need thinning more than Compt. 8 does. Power to exercise discretion in such matters should be prescribed in the plan, and often the extent to which officers of various ranks may sanction minor deviations is specified. For instance, alteration of thinning years might be at the discretion of a District Officer, of thinning-cycles at the discretion of a Divisional Officer, but alterations in the prescribed methods of thinning might need the sanction of a Chief Conservator.

Another method of approach suitable for areas in which it is known that skilled and reliable staff will always be available, is to prescribe that localities listed in the tables for the various operations in each year shall be inspected in the year stated, with a view to deciding whether the operation is needed at that time or not. If it is needed it will be done and recorded as such, but if not, it will be carried forward in an arrears column of the table for periodic inspection until it is done.

When the location of all exploitation operations has not been prescribed definitely in the plan, columns must be provided in the tables of cultural operations in which the sites of any felling done will be entered as soon as it takes place, so as to bring the areas into the tending schemes. In this way proper attention will be assured, for instance, for groups of advance regeneration which have been released.

A separate chapter having been written for the Special Plan for each Working Circle prescribed, the remaining chapters of the plan will deal with the whole working plan area.

MISCELLANEOUS PLANS

This chapter is used for the prescription of any work which is required anywhere in the working plan area which has not already been dealt with in the special plan for any working circle. A fairly wide selection of operations is suggested in the format and there is often a choice between prescribing certain operations in the special plans or in this chapter.

Fire danger, for instance, may be a major factor over the whole area, calling for a special fire plan, with look-out towers, special postings of staff at certain dangerous times of year, telephone lines, trucks, and other equipment, or may be merely a local factor affecting a few compartments beside a railway during dry seasons only. In the latter case the simple prescriptions needed will have been included in the cultural prescriptions of Section 6 of the special plan for the circle concerned. In the former case it will form one of the miscellaneous plans set out in this chapter. Similarly with drainage and maintenance of boundaries.

It is in this chapter that many of the possible improvements which the working plans officer would like to effect are dealt with.

Improvement of Communications, for instance. Good extraction roads serving all parts of the forest, so arranged that only short downhill hauls are needed to get produce on to surfaces on which transport can move them economically, may make all the difference between economic and uneconomic forestry and will certainly make possible more intensive working. The availability of extraction routes has a profound effect on management. Where there is a strong demand for all sizes of timber, poles, and fuel it is possible to sell small quantities from any part of a forest which is divided into a number of small compartments by a network of well-graded roads and rides, at any time the manager wishes. Therefore a plan can be made for the harvesting of frequent light thinnings and of small numbers of trees as their increment rate culminates, wherever they are in the forest. The best possible use can be made of the soil, advantage can be taken of favourable markets and of good seed years, poor trees can be eliminated quickly and all waste can be avoided. In a large tract of forest at a distance from markets and without existing roads a considerable volume from one portion of the area will be needed to cover the expense of making a route to extract it. Immature as well as mature trees may have to be cut, and the location of the subsequent fellings may be determined more by the convenience of extending the road or putting in feeder roads to it than by the silvicultural needs of the various parts of the forest. Roads and markets are linked in a vicious circle. With a good market it pays to put in plenty of good roads, and then any produce anywhere can be sold. With a poor market few roads can be afforded and much growth will be wasted.

Nevertheless, bold investment in extraction routes, be they roads, tramways, or cables, will generally pay in the end, and will certainly make better management possible. Roads, too, are of great value in fire fighting, enabling men and equipment to be rushed to an outbreak before it spreads, as well as providing cleared lines from which operations can be started. They are, however, expensive, particularly in hilly country, where numerous bridges are necessary. To budget for putting in all the roads required in a forest during the first period of a plan would involve the owner in very considerable capital expenditure which he would not be prepared to face. Prescriptions for the gradual building up of an adequate road-system out of each year's profits would be much more acceptable. The important roads will be those on which the cut during the working plan period will be brought out. In forests where roads already exist, improvement of their surface may be all that is called for, and in any case provision will have to be made for maintenance. A programme of metalling or re-metalling definite stretches each year, probably before the annual extraction if this is seasonal, and of the repair of all road drains damaged during extraction before water erosion becomes serious will be necessary.

In some countries where forestry is in its early stages it is customary to leave extraction entirely to exploiters. The main reason for this is that the staffs of the forest departments are numerically inadequate for all the work there is to do. Where reservation has not yet been completed, survey and demarcation are still going on, and the composition and regeneration possibilities of most forests are unknown, foresters cannot be spared to build roads and contractors are unobtainable. The best that can be done is to retain control of where the roads shall be made. One African working plan contains the following prescription: 'Main extraction roads shall be laid out in an orderly manner, their alignment being planned by the executive officer (i.e. the district forest officer), with the co-operation of the licensees, who will bear all expenses of construction.' Naturally, as the licensee has the expense the rates he pays for the timber are lower than they would be otherwise, but even if he paid more it would not always be possible to find the staff to make the roads.

Although stress has been laid on roads in what has been written above, the possibilities of other means of extraction should not be overlooked. For instance, modern mechanically driven cableways are being installed in many mountain forests, including several in Africa, and they are particularly suitable for use in plantations. In plantations, too, extraction damage can often be minimized by the prescription of temporary extraction routes in the form of line thinnings wide enough for a cart or light tractor, spaced at suitable distances apart.

Another matter which may need special planning in a newly opened forest is accommodation for staff and labour. In this case, too, it might be very expensive to do all that is required at once, and it may be necessary to build temporary houses, labour camps, etc., to be replaced by permanent structures on a planned programme.

When a first plan for an area about which little is known is being drawn up, it is in this chapter that prescriptions are made for doing during the period of the plan those things which will lead to the preparation of a more adequate plan in the future. Prescriptions for additional survey and enumeration, for completion of stock-mapping, for experiments in various methods of obtaining or tending regeneration, for the establishment and periodic measurement of sample plots, for the study of rates of growth, etc., can be made here.

ESTABLISHMENT AND LABOUR

The existing state of staff and labour has been set out in Part I. If any alterations will have to be made in numbers, organization, or methods of employment they should be prescribed after a brief statement of the reasons. Additional staff and labour could almost always be used to the advantage of a forest, but what should be stated here are the minimum numbers required to carry out the operations prescribed if efficiently organized, and any steps which may have to be taken to obtain them. If additional housing or labour camps will be needed the necessary prescriptions will have to be made in a separate section.

Probable costs of the staff and labour for the period of the plan, including any necessary increments or alterations in piece rates, must be given. Labour costs may be distributed among the estimates of the various operations to be carried out, or sometimes an adequate number of regularly employed labourers may be prescribed for all the works required and their wages shown as a lump sum.

CHAPTER

16

CONTROL OF THE PLAN

It is useless to make plans if they are not to be carried out: hence, the system of sanction by a high authority, Minister of State in France, Governor in a colony, etc., under which the delaying action of red tape provides a safeguard against any hurried or improperly considered attempt at alteration. As already mentioned, powers should be incorporated in the sanctioned plan for minor changes, adoption of improved techniques, etc., to be authorized by departmental officers.

The idea of sanction and control is not to clamp an out-moded form of management on to a forest after it is known that something better can be done, but to ensure that a carefully considered scheme is given a fair chance. A working plan should embody the agreed conclusions of all those best qualified to suggest the manner in which a forest should be managed at the time at which the plan is drawn up. It is, therefore, important that it should be carried out as intended until there is clear evidence that there are flaws in it. Any forester who believes that he has found such a flaw and that he knows of an improvement which cannot be adopted within the framework of the plan, should draw attention to it. If he can convince the officers concerned with the administration of the area, the plan should be amended or revised and the alterations sanctioned. This procedure will cause some delay, but it will prevent an individualist who thinks he knows best from discarding lightly a scheme which others have decided should be tried, and will make it necessary for him to draw up a reasoned case for his own views.

Actually the weaknesses of a plan are most likely to be shown up by the results obtained in working to it. Steps must, therefore, be taken to check that prescriptions are being carried out according to the schedules and to focus attention on any points at which progress is falling into arrears or running ahead of the plan. Examination of such deviations will show whether the fault lies in the plan or in its execution. This is what is known as control of a plan. The machinery of control in state-managed forests will work somewhat as follows. Each year when the district officer is making his expenditure estimates for the coming year he studies the operations prescribed for that year. If, in the light of his knowledge of the state of the crops in the various compartments, he thinks that some of them should be altered, he will make any such alterations as are within the powers conferred upon him. He will then put up proposals for any of the alterations which are beyond his powers to his senior officer. The latter will probably make an inspection and if he agrees will either give or obtain written sanction. At the end of the forest year the work actually done and the results obtained are entered in blank columns added for that purpose to the tables of operations prescribed. Remarks explaining discrepancies or quoting written sanctions obtained are added. These entries are actually summaries of the work done and the results obtained in individual compartments, as recorded in the Compartment Registers when each operation is carried out (see page 76).

Office work is one of the bugbears of all organizations, but a certain amount is essential for efficient operation, particularly in long-term projects, such as forestry, and maintenance of adequate records is an essential part of management. Many forest departments have their own standardized forms for record and control, devised so that essential information passes with a minimum of work from daily labour tickets, or whatever the system for payment of labour is, to monthly records, annual records, and control forms. Working plan procedure should not increase office work, but rather replace some of it, simplify and reduce such items as annual estimates and correspondence about the area.

When standard control forms are to be used in a department all that is needed in this section of the plan is a prescription that certain specified forms will be completed by a certain officer and sent by a certain date each year to a specified destination. Often this is done by means of a duplicate known as a flying copy, which is sent to, say, the Conservator of Working Plans, who checks it, has his own office copy brought up to date, and returns the flying copy for use again next year.

When there are no standard forms suitable for the particular plan on hand, the simplest possible forms to meet the needs of the case should be drawn up. Often the addition of columns to the tables of operations, i.e. the plan of exploitation (page 141), the plan of cultural operations (page 144), etc., to show the year in which the operation is actually carried out, the area completed, or the volume extracted will meet the case. A column for arrears or deficit, or of excess over the amount prescribed will be useful, as also will be space for remarks.

A common arrangement for control of volume output is:

Year.	Cpt.	Species.	Prescribed output, cu. ft.	Actual output, cu. ft.	Excess (red) or Deficit (black), cu. ft.	Running Excess (red) or Deficit (black).	Remarks.
1950	2	Oak	20,000	22,000	2,000	2,000	
		Beech	12,000	11,000	1,000	1,000	
1951	9	Oak	15,000	14,000	1,000	1,000	
		Beech	10,000	11,000	1,000	—	

For regeneration by simple planting a form showing progress, with provision for carrying forward any arrears not successfully started each year, is needed, e.g.:

Year.	To be planted. Cpt.	Area.	Attempted. Cpt.	Area.	Successful. Cpt.	Area.	Arrears to be carried forward. Cpt.	Area.
1950 Arrears .	—	—	—	—	—	—	—	—
New prescpt. .	1	100	1	100	1	75	1	25
Total . .								
1951 Arrears .	1	25	1	25	1	25	—	—
New prescpt. .	2	100	2	90	2	80	2	20
Total . .		125		115		105	2	20

For natural regeneration columns are needed to show the area under regeneration which must be carried forward each year and the area on which regeneration can be considered to be completed.

For Miscellaneous Prescriptions a very simple form, as follows, can be used:

Year	Prescriptions Operation and Locality prescribed.	Year(s)	Action taken in which carried out.	Remarks
1950	Repair 200 yards of boundary fence north side of C. 5.	1950	Fence repaired, 200 yards.	

Sometimes a Record of Deviations from Working Plan Prescriptions is kept in a simple form, showing for each year what prescriptions have not been carried out fully, and quoting the authority for each deviation.

Record forms other than control forms will generally be needed in which running records of operations, such as exploitation and regeneration and expenditure, under various heads, will be summarized from the Compartment Registers. These will be standardized in most Forest Services and coordinated with the forms required as appendices to annual reports.

In France records and control are combined for each series in a Management Notebook about $9\frac{1}{2}$ inches by 6 inches, bound in stiff covers and convenient for carrying in the field. The detailed lay-out of the book varies for each Conservancy and the standard headings, etc., adopted for a region may be altered to suit the local requirements of a particular forest or plan. The notebook always opens with a brief statement of the name, area, and particulars of the forest, the composition of the series, and extracts from the official decree, setting out the method of management and the yield sanctioned. The plan of exploitation is given in tabular form, showing the year, location, and area of each regeneration and tending felling, followed by the cutting rules for each type of operation. A table of the growing stock of each compartment divided according to species, age- or size-classes, as appropriate, is followed by the details of the yield calculation, whether by volume or area, and by the volume tables used in the enumeration. Then there is a closely-ruled page for each compartment, on which are recorded the numbers of trees and the volumes removed each

154

time a compartment is entered, differentiated according to the class of felling carried out and with special columns for windfalls. Pages are also ruled for annual summaries of exploitation of all kinds and for windfalls in the regeneration and tending areas respectively. Finally, tables are provided for annual summaries of all other work carried out, maintenance of boundaries, roads and rides, cultural operations, etc., with a wide column for notes on any form of damage, tenders for contracts, and on 'all matters affecting the forest'. Room is also available for general inspection notes and comments.

Four copies of each book are kept, one by the administrative headquarters in Paris, one by the Conservancy, one by the Division (Inspectorate), and one by the District. The district copy is kept up to date by the local officer and is sent up once a year so that the others may be completed. This provides a very compact and comprehensive method of recording the progress of all operations, the working of the plan, and information which will assist the revising officer in due course.

CHAPTER
17

ESTIMATE OF THE CAPITAL VALUE OF FOREST

Often this cannot be made with any reasonable accuracy, and for that reason it is omitted in many plans. However, if a fair estimate of the growing stock has been made and the local value of land of similar quality is known, an approximate figure can be given which will show the order of magnitude of the capital value of the forest for which management has been planned.

FINANCIAL FORECAST

The volume of the annual out-turn of all classes of produce has been prescribed or estimated and the work to be done each year has been prescribed. By assigning anticipated values to the produce and estimated costs to the operations a balance sheet can be drawn up which will show the net annual surplus or deficit during the period of the plan. The tables of cutting and other operations provide the material for the calculations, and as already suggested columns for values which it is expected will be realized for the cuts shown in the tables, and for the anticipated cost of each operation can be added to the tables.

If the capital value of the forest has been ascertained it is possible to express a net surplus as an annual interest percentage on the capital.

COST OF PREPARATION OF THE PLAN

If the plan has been drawn up by a special working plan party the cost of its preparation should be stated here. If it has been made by local staff in the course of their normal duties it is often impossible to assign any special figure to it.

APPENDICES

The question of the distribution of data between the text

of a plan and its appendices has already been discussed (page 57). In general, anything in the nature of detailed lists of figures extending to more than one page should be relegated to an appendix, and a brief summary or appreciation of them should appear in the text.

SUMMARY OF PRINCIPAL PRESCRIPTIONS

At the beginning of the layout of a plan attention was drawn to the advisability of preparing a summary of the Principal Prescriptions of the plan, which is really an appendix best placed at the beginning of a plan. It can, of course, take many forms, either narrative or tabular. No one form can be suggested as the best to cover all types of plan, and the aim should be to give the maximum of information in the most concise manner possible. In some cases, particularly when it is desirable to supply simply expressed instructions for subordinate staff with little technical education, the summary of principal prescriptions can be supplemented by an annual table of operations prescribed for each year, showing the compartments and extent of each operation. Various types of summary are to be found in different published working plans.

A summary for an imaginary case might run as follows:

Ref. para. no.
of Plan or pre-
scription no.

3	*Area*	Productive	.	10,000 acres
		Unproductive	.	120 ,,
		Total	. .	10,120 acres

40   *Division into Working Circles*
        (A) Hardwood Working Circle   4,000 acres
        (B) Coniferous   ,,   ,,   5,900 ,,
        (C) Amenity   ,,   ,,   100 ,,

45   *Period of the Plan* 1st October, 1950—30th September, 1959.
        (A) Hardwood Working Circle.

48   *Felling Series* One only.

50   *Species and Rotations* Oak and Beech 150 years.
                          Ash and Sycamore 75 years.

51   *Silvicultural System* Uniform with natural regeneration.

55   *Regulation of the Yield* By volume, based on size groups.

Ref. para. no.
of Plan or pre-
scription no.

56	Current regeneration block, 1,000 acres of woods, 130 years and older.
57	Thinning block, 3,000 acres of younger woods.
60	Average annual yield of regeneration fellings 100,000 H. ft.
62	But seeding fellings to be accelerated before and immediately after mast years.
63	Thinning-cycle 10 years to cover 300 acres a year. Average annual intermediate yield estimated 30,000 H. ft.
	(B) Coniferous Working Circle.
	Felling Series.
	Species, system, yields, etc.
	(C) Amenity Working Circle.
	Species, system, etc.

Miscellaneous Operations

110	*Roads* Annual plan of resurfacing about a mile of main extraction road a year.
112	*Fire Protection* Approximately 10 miles of firebreaks to be cleaned by 1st April each year.
113	Fire towers to be manned in accordance with standing instructions.
	Financial Forecast

		£
120	Estimated annual revenue . .	
	,, ,, expenditure .	
		───
	,, ,, surplus . . £	
		───

Compartment Descriptions are the most important of the appendices. All other information collected, such as descriptions of soil pits examined, enumeration survey data, increment study data, etc., should be placed in appendices for permanent record and for the use of the officer who will revise the plan. Use can often be made of graphs as a means of presenting data.

MAPS

An index map is one showing primarily topographical features, roads and rides, fire lines, drainage system, and division of the area into compartments. Allocation to working circles, felling series, etc., can sometimes be shown on the index map, or sometimes a separate management map is prepared, which can also be used to record the progress of operations, such as exploitation and regeneration, by distinctive hatching.

Geological and soil maps should be combined whenever possible to show the distribution of sites, the differences between which have been recognized as having practical effects on management. A stock map is important and has already been discussed.

A fire control map will only be needed when fire is a serious menace to the area, necessitating a special fire-plan.

The scale of maps should be suitable for the intensity of working which is planned, but foresters often have to make do with what is available. Except for small areas maps on a scale such as 6 inches to 1 mile, 1/10,000, etc., are generally large enough. For large areas, to be worked extensively, something like 1 inch to 1 mile, 1/50,000, etc., may have to be adopted. At the other extreme, maps at 25 inches to 1 mile are sometimes used.

CHAPTER
18

The form of a working plan should be adapted to suit the conditions of the tract for which it is being prepared. Attention has already been drawn to the fact that plans have to be made for the management of a variety of forms of vegetation, of areas in very different stages of development, and for the attainment of different objectives. The list of headings suggested for a plan has been given with the warning that all the sections will not be required in every case and that additional sections may be needed in others. A few examples will make this clear.

PLAN FOR A PROTECTION FOREST
In the very early stages of the application of a forest policy to a country in which little development has taken place it is often necessary to constitute some forests primarily for protective purposes, for instance, to prevent temporary farming from causing floods and soil erosion off the steep slopes of a water catchment area. Normally there should be no rigid differentiation of forests into areas for protection and areas for production, because all forests should exercise protective influences and at the same time produce continuous crops of value to the community. It must often happen that one of the two functions is of greater importance locally than the other even in well-developed countries. Many tropical rivers which water heavily populated districts have their catchments in remote regions where sparse and primitive populations have little need of forest products and where there are no communications. In such conditions protection forests have to be constituted and little production will be possible from them for many years.

The most important points of a management plan for such a forest are statements of situation, distribution, and legal position, including a careful record of any rights admitted, descriptions of the boundaries, and prescriptions for their maintenance, for prevention of encroachment, and of damage by fire. As in every other case in which a first working plan is being made for an area it is wise to investigate and record the history as far back as it can be ascertained, because revising officers will always assume that this has been done, that all possible sources have been tapped and that there is no point in making further researches. The history will often throw considerable light on the present state of the tract and its vegetation, and knowledge of it may prevent wrong conclusions from being drawn. If any steps are to be taken to improve the water-regulating properties of the tract, then the topography, geology, soils, climate, ecology, and local fauna will have to be studied, but without a market for produce it would be difficult and expensive to carry out any treatment. In the conditions of inaccessibility and lack of population envisaged the first plan would probably be a very simple one. The object would be the prevention of destruction of the vegetation by man and fire, with the hope that such protection would allow the soil cover to become more complete. Hence, Part I of the plan would consist of Chapters 1 and 3 of the suggested format, followed by the objects of management and prescriptions for protection in Part II. A plan of communications to make the interior accessible for fire-fighting might be necessary. A special prescription might well be included to the effect that, whenever opportunity arose, the local staff would collect further information about the forest and record it in the plan file. Possible marketable products, the scale and manner of extraction which might become possible, regeneration, and growth would be studied from time to time on tours of inspection, and noted in readiness for the day when increase of population or the opening of new communications might make it practical to organize production. Such studies have often been made in similar circumstances in the past, but the results were generally buried in monthly reports or in the files of miscellaneous correspondence and so were not readily available when needed. The adoption of the working plan file as the focus for recording all that is learnt about a forest is not only a matter of practical convenience, but the fact that there is

F   161 such a file will stimulate further study of the area by succeeding district officers.

It is sometimes necessary to create a source of supply of forest produce, either for a particular need when no natural source exists, or as a supplement to inadequate natural provision. The material wanted may be firewood, a minor forest product, timber, or a combination of these. Ideally a series or a number of series should be created by equal annual plantings of the desired size for a number of years equal to the proposed rotation of the new crop.

Often speed is an important objective and planting proceeds more rapidly with the intention of making available a large volume of thinnings and some premature main fellings before the end of the rotation. Though this is legitimate it should be planned and the pace of establishment should not be left to chance and the annual vagaries of fluctuating funds and labour.

As every afforestation scheme is an investment of capital intended to create a new and permanent source of supply of raw material, it is particularly important that it should be carefully organized. The first requirement is an area of soil suitable for growing the type of produce wanted, of such an extent that the species it is proposed to establish will be able to provide a sustained annual yield of the required volume on the sites available under the existing climatic conditions. This calls for full investigation of the legal position, configuration, soil types, and climate of the suggested tract. It is assumed that trials of the species to be used will have been made before any large scheme is launched, or that adequate evidence exists that they can be grown there satisfactorily. The size of area required may be determined by the need to provide an annual output sufficient in itself to keep a sawmill or other processing plant of economic capacity constantly provided with raw material. There have been cases in which this was the object, but all the available planting sites in the vicinity were planted up before it was achieved, and isolated incomplete series of little economic value resulted. This could not have happened with proper planning, and it is essential that the first chapter of Part I of a planting plan shall record the acquisition of land adequate for the completed project, and the settlement of all rights on it. The studies of local conditions necessary for completing Chapter 2 need to be carried out very carefully, and the history of the tract will often explain the state of the present vegetation and the soils which have formed under past usages. Ecological investigations will complete the information necessary for the allocation of the various sites to the species likely to grow best on them. Economic considerations are the reason for the scheme and special attention will have to be paid to communications and their improvement, but the section of most immediate importance in the fifth chapter will be that dealing with the labour supply. When any afforestation is undertaken the number of labourers needed increases very rapidly as areas to be tended and thinned accumulate, and many schemes have failed because they did not take this into account. Careful estimates of the numbers required after several years for new planting, tending, thinning, road making, etc., and of the potential numbers likely to be available locally or obtainable from other districts on contract, etc., must be made. The suitability of the labour for the types of work required must be considered as well as their numbers and capacity. Only when all this has been done can the scale and pace of establishment be decided. In Part II of a planting plan the objects of the scheme will be stated clearly and the general plan outlined before the detailed work for the first ten years is prescribed. Particularly important will be the principles to be followed in laying out internal communications and compartments. For both these a planting plan provides a wonderful opportunity to start off on the right lines. Presumably there is nothing on the ground which is not going to be cleared, and both contours and soil boundaries can be studied with a view to a convenient network of roads and rides which will tap all compartments and provide permanent boundaries for them. These compartments can be as nearly as possible homogeneous in site and of a size most likely to facilitate management. Possibly a sawmill or other processing site will have to be selected and reserved, in which case the road system will be designed to feed it. The roads will certainly not all be made up at the beginning, but the main routes at least should be surveyed and probably pegged out before the area is obscured by trees. The general scheme of minor roads and rides to serve as compartment boundaries should be settled, and those round the compartments to be planted in the first ten years should be mapped. Some foresters maintain that it is wasteful to leave unplanted road and ride sites which will not be needed until there are thinnings to be extracted, but this overlooks the fact that men, tools, and plants have to reach the planting areas, and that rapid access to all planted areas in case of fire may be essential. When the expense of extracting the planted stumps from the ride sites is taken into consideration there is often little in the economic argument. At the same time as the communications are considered the scheme of fire protection will have to be worked out, so that those roads which will form part of it can be supplemented by belts of fire-resistant species or by cleared traces, and any other belts or traces required which will not be based on roads can be mapped.

Any division of the area into working circles for the different types of forest to be created, and their sub-division into felling series of convenient sizes for annual planting areas or for the eventual supply by different main extraction routes will have to be prescribed. For each circle the species, methods of establishment, and tending will be outlined, and detailed tables of operations for the first ten years will be drawn up. Special plans will probably be needed for road construction, fire protection, drainage, building, and nursery work, with the general scheme for each and the details for the working plan period. Thinning may or may not have to be provided for, depending on rates of growth and the period of the plan, but in any case it will have an important part in later revisions.

The details of recording and control of operations will need particularly careful organization and prescription. The opportunity of providing full information from the beginning about the building up a new forest must not be missed. How fortunate the foresters who will manage the area in the future will be when they do not have to guess how the various stands were established, what proportions of species were planted, how they were thinned, and so on. Nevertheless, care must be taken not to create too much office work, and the minimum number of forms should be designed to suit the staff available, the accounting methods used, etc. Compartment registers can be started one by one as each compartment is laid out and entered.

PLANNING FOR AREAS OF ABNORMAL AGE-CLASS DISTRIBUTION

As has been mentioned above, plantations have often been established as rapidly as possible in order to provide some produce in the shortest possible period. Times of trade depression have sometimes curtailed planting funds, or in some countries have lead to the use of unemployed persons on accelerated planting, and wars have upset all calculations. The result is that foresters are often faced with the organization of management of blocks of plantations with very abnormal age-class distributions, sometimes with portions of natural forest or open land which can be included in a working plan area, and sometimes without. The problems of planning are sometimes not very different from those of a previously unmanaged natural forest, except that it is the younger age classes which are excessive and the mature ones missing, instead of the other way round. The aim will be to work gradually towards increasing a probably non-existent final yield to the sustained yield capacity of the tract as the plantations mature. This will generally involve the regeneration or replacement of the least satisfactory trees or stands before they reach maturity or the intended rotation. It may be possible to use such fellings to prevent great fluctuations in the output which will be mainly dependent on thinnings. The organization of thinnings will frequently be the most important part of the plan, because it often happens that when plantations have been created in a hurry, or a war has intervened, thinning operations have fallen seriously into arrears. Even if additional bare land is available for planting in order to improve the distribution of the age-classes, further planting may have to be postponed until a normal thinning-cycle has been achieved. Variation in thinning density may be adopted to help to adjust the size-class distribution as a substitute for age-class distribution. For instance, some stands may be thinned heavily to stimulate diameter growth and so provide for premature final felling.

Sometimes it may be possible to group several isolated areas into the beginning of one felling series, while in other cases large planted tracts can be divided up into a number of potential series for the convenience of management. Internal communications are likely to be unsatisfactory and a system of roads and rides may be needed urgently for the extraction of thinnings. Frequently, when dealing with previously unplanned plantations, all sections of a normal working plan will have to be written except that concerned with the final yield. If previous records are non-existent or of little value there will be no point in retaining any previous system of compartments, and new ones can be laid out to suit the management which is to be provided.

REVISION OF PLANS

Every plan should prescribe the date, generally a year before the expiry of its period, at which the collection of data for its revision should be started. Actually, of course, such data are being collected continuously under a working plan, but the putting of the figures into order and the reassessment of the growing stock are part of the revision of the plan, which must be completed in time for new prescriptions to be written and approved before the original period expires. Even before the time for revision arrives in its normal course it may become apparent that the existing plan is not working properly. Regeneration may be falling behind schedule, final fellings may have extended over a larger area than that expected to provide the prescribed yield, or sufficient labour may not be available for all the operations scheduled. In such a case the plan or some part of it will have to be revised at once.

The standing instructions of the French forest service include special provision for revision of the yield regulation of a plan only, as well as for normal revision.

If an original plan has been drawn up well its revision should be a fairly simple matter. There is no need to repeat the descriptive matter of Part I and the appropriate sections of the revision can be used to record any corrections or additions. The most important section will be that bringing the history up to date by a detailed examination and a review of the results of the management during the period which has just ended, including a summary of the yields and financial results obtained. A new Chapter 6, 'Statistics of Growing Stock and Increment', will be provided when fresh enumeration data are available, as they should normally be for a revision. If sufficient reliance can be placed on the two sets of figures and on the records of removals the growth of the stock during the period can be calculated.

If the objects of management and the general scheme to attain them are to remain unchanged, the revision of Part II will consist primarily in a fresh prescription of yield and in the preparation of the detailed tables of operations to be carried out in each working circle during the coming period. Prescription of techniques of regeneration, thinning, felling, etc., may be revised in the light of experience gained. Some alterations may be needed in the miscellaneous plans for communications, fire protection, etc., and new schedules will be needed for their immediate progress. Special attention should be paid to the manner in which the machinery for records and control of operations has worked, and any modifications or improvements suggested by the experience gained should be prescribed. In the final chapter a forecast of revenue and expenditure for the new period will be required.

It may be, of course, that some drastic alterations are necessary, for instance, the adoption of a different silvicultural system, revision of the working circles, or of compartments, though this last should be avoided as far as possible because of its effect on records.

The main value of a working plan lies in progressive accumulation of knowledge about an area and its growing stock, so that each revision should have a sounder basis for its prescriptions. The author of an original plan can only estimate what he thinks can be done and what the results will be. The revising officer has before him on paper and on the ground the actual results of the prescriptions made some years before, and from careful analysis and examination of these he should be able to decide what can be done in the next period. He will pay particular attention to any points at which achievement during the first period fell short of the expectations of the planning officer, and should try to ascertain whether this was due to faulty operation or over-optimistic expectation.

One of the most masterly appreciations of the spirit in which a working plan officer should set about the organization of a forest was written by Mons. Broillard, Professor of Forest Economy in the School of Forestry, Nancy, in his *Cours d'Amenagement des Forêts*, 1878 (translation by E. E. Fernandes, of the Indian Forest Service, slightly amended).

'Every forest offers a real and living individuality. It differs from every other forest by its situation, its aspect, and configuration; by its soil, by its component crops, and also by the character of the surrounding country. There are no two forests any more than two towns exactly alike and it would be a great mistake to suppose that the management of forests adjoining each other or situated in the same region can be built up on the same framework or pattern. The forester labouring under so erroneous an impression would lack the very fundamental idea that should guide him, and instead of adapting himself to circumstances would vainly endeavour to force circumstances to suit his silly imaginings.

'The great dangers to be avoided in forest management are preconceived ideas and foregone conclusions. Every rigid system refusing to yield to the varying requirements of different forests and localities must be equally vicious, and more than this, it must infallibly result in its slaves overlooking some important facts and indispensable conditions. Indeed, it is this very danger of carrying into effect preconceived opinions that justifies us in warning the forester against seeking any perfect solution of the problem before him, the realization of an impossible ideal, and in advising him to confine himself to doing his best to obtain the results required and no more. If, embued with this spirit, he knows the forest he is dealing with, is careful to conform to the essential rules of forest management, and allows himself to be guided by the true principles of silviculture, by endeavouring to obtain from well-constituted crops and promising trees only such products as the soil can yield, he will scarcely ever fail to draw up a good working plan.'

PART III

THE DEVELOPMENT OF FOREST MANAGEMENT IN WESTERN EUROPE

CHAPTER
19

The history of natural forests has followed a very similar pattern practically all over the world, up to a certain point.

In the climates of those parts of the earth destined to carry considerable populations the natural vegetation of large proportions of the land was forest. This forest was first regarded as a subsidiary and casual source of food from the wild game and fruits found in it, and as a refuge in times of attack by superior numbers of hostile neighbours. Then as populations expanded, forest was an obstacle to production of food, either by arable farming or the herding of domesticated animals. Trees were admittedly a source of heat for cooking and comfort, and produced material for housing, implements, etc., but there were so many of them that those felled to make room for crops provided all the wood that was needed. Gradually this ceased to be the case and wood, particularly for fuel, had to be sought further and further afield. What was wanted was cut, the poorer trees and the less useful species were left, and gradually large areas of forest, uncleared because they were not needed or were too steep for farming, were so impoverished as not to be able to supply the wood needed. This lead to the imposition of restrictions on cutting, but as little was understood about the growth of forests and practically noth- was done to assist it, the restrictions merely slowed down destruction and annoyed everyone. Industries such as iron smelting developed and increased the demand for wood, communications improved, making further sources available, and the growing power of money strengthened the resistance to restrictions. Efforts began to be made by a few people to spread the idea that forests should be cropped and tended instead of being mined, but this did not suit the people who were making money. They invoked financial arguments that it was essential that their operations should be unfettered and public opinion was on their side. Only when so much of the natural forest had been destroyed or ruined that the economic life of the country was in serious danger were the public ready to accept restrictions, this time accompanied by efforts to rebuild and extend the remaining forests. Foresters were then allowed to take charge of forests and to attempt the beginning of management and silviculture.

This point was reached in different parts of the continents of Europe, America, Australia, Asia, and Africa at various times from the end of the 17th century to only yesterday. As it happened first in parts of Western Europe it is useful to trace the development of forest management in these parts from its crude beginnings to its current practice and probable future trends.

WESTERN EUROPE

During the first centuries after Christ, primeval forest covered much the greater part of Europe and was something that had to be cleared before land could be put to profitable use, just as it was in North America and Australia at a much later date. The requirements of the local population for fuel and timber were almost always easily met from new clearings for agriculture, and when they were not, haphazard felling could be carried out at the most convenient place. The earliest written record of any management for forest produce is contained in Pliny's 17th Book (about A.D. 50), where he tells us that in the southern wine-growing districts of ancient Gaul Spanish chestnut was cultivated on a three-to five-year coppice rotation for vine stakes. He also says that where people had to go a long way for fuel, oak was coppiced on an eleven-year rotation. This state of affairs continued practically throughout the Feudal Period, between the 10th and 14th centuries, but clearing for settlement increased apace towards the end of this time, and some measure of control became necessary. The first measure adopted in France was that in populous districts forest cutting had to be carried out in adjoining areas from year to year, a process known as *couper à tire*. The area which could be cut each year was not restricted, but in each coupe the retention of some vigorous seedling trees of oak and beech was ordered, with the intention that

172

they should grow on as standards and seedbearers over the coppice. At that time fire-wood was of great importance, and in France its production from coppice, together with some timber from standards over it, became general.

Severe penalties were ordered for the wanton destruction of forest by fire, and for the grazing of cattle in the coppiced coupes. Grazing and pannage of swine in high forest were also restricted, largely to the herds of noblemen.

Over much of the zone affected by these orders beech was the natural dominant, but the coppicing favoured oak and ash, which reshoot more freely, and also ash and birch, which seed at a relatively early age. On account of the greater durability of its timber oak standards were preferred for retention as timber trees, so gradually the dominance of beech was reduced.

From about the middle of the 14th century, both in France and in the various German principalities and states, the need to restrict the size of annual coupes in order to make forests last until the first coupe which had been coppiced was ready to cut again, was felt. This control by area (*couper à aire*) of equal, adjoining (*couper à tire*) annual coupes came to be known as the method of *tire et aire*. It was applied to coppice on rotations of up to twenty years with, in many forests, particularly in France, the retention of seedling trees of rotation, twice, and thrice rotation age, thus confirming the practice of Coppice with Standards. *Tire et aire* was also applied to those parts of the accessible high forests of oak and beech which were not being coppiced, rotations of 100 years and more being adopted. The *Ordonnance de Melun* of 1376, which has been described as the first French forest code, prescribed that area coupes should be of 25–40 acres in high forest and that they should be closed against domestic animals. In these coupes some 5–10 seedbearers were to be retained per acre, and some sowing, particularly of oak, was carried out to supplement natural regeneration. In the earlier stages of their growth these coupes had the appearance of coppice with standards, but later they developed into two-storied high forest, that is to say when they were successful. Actually they were very rarely fully stocked, nor did they often contain many of the desired species, oak and beech. More often hornbeam, ash, cherry, wych elm, yew, alder, aspen, poplar, and birch predominated. However, it was in these coupes, when the young crops were dense, that the need for cleaning and thinning was recognized and the first tending operations were practised locally.

On the lower slopes of the Pyrenees coppicing of beechwoods over considerable areas caused serious flooding and erosion, and the practice of selection coppice, under which the soil was never completely exposed, was evolved.

Nursery work was unknown at this time, but it gradually became general practice to sow acorns and beechnuts, or to transplant several natural seedlings wherever gaps were made by felling mature trees. In 1368 the city of Nuremburg, in North Bavaria, sowed Scots pine, silver fir, and Norway spruce seed on a large area which had been burnt near the city. This was so successful that in 1420 the city of Frankfurt-on-Main obtained Scots pine seed from Nuremburg and sowed it on waste areas of considerable size on the sands and gravels of the Rhine-Main plains.

From about the middle of the 16th century, earlier in some places and later in others, the demand for firewood and timber began greatly to exceed supplies from local sources. Firewood and charcoal were the only fuels, and enormous quantities were consumed in the salt-works, forges, and glass-works that had sprung up. So acute was the shortage of firewood that in some districts communal baking ovens had to be installed to economize its use. Water-powered sawmills had become highly profitable, floating was started on many continental rivers, and as prices rose forests further and further from the centres of consumption were exploited, often ruthlessly, with no thought of the future. Oppression of the peasants led to much felling and forest grazing in excess of the rights granted, and the disturbance caused by wars, the Thirty Years' War in Germany (1616–48) and the Civil War in England (1642–46) led to further destruction. The virgin forests of Western Europe, including those of the British Isles, were reduced to inaccessible blocks on the mountains, and by the middle of the 17th century it is fair to say that the rest of the natural forest heritage had been destroyed. For lack of renewal of the stools after repeated cutting most of the coppice areas were moribund. In the high forests almost all the utilizable timber had been felled, leaving a few scattered, stag-headed oak and beech, with occasional groups of promising poles, and the rest a waste of weed species, brambles, bracken, and heather.

Most of the Western European forests to-day are essentially the work of foresters during the last two or three hundred years.

In France the great administrator Colbert appointed a Commission in 1662, and as a result of the work of the energetic men who composed it a new Forest Code was issued in 1669. This applied not only to State forests but to all forests, irrespective of ownership. Detailed regulations were concerned mainly with hardwood forests. Minimum rotations were laid down for coppice and for high forest, a minimum number of seed-bearers had to be reserved per acre, and all blanks had to be sown with acorns. Moribund coppice had to be rejuvenated by burning the brush piled on top of the cut stools in order to stimulate the production of shoots from below ground and the consequent formation of new roots. This practice, known as *sartage* when combined with harvesting a crop of rye from the year's coupe, had originally been carried out to supplement local food supplies in the poor hill country of the Ardennes, but its rejuvenating effect on coppice had made it a forestry measure. Young high forest crops of poor growth were ordered to be cut back at 20–25 years of age and allowed to reshoot. The importance of thinnings and cleanings in young crops was stressed.

In England John Evelyn published his *Silva* in 1662, but failed to arouse the public conscience. It did, however, stimulate one of the periodic waves of activity in the New Forest.

In France in 1727 clear felling was prohibited in the remaining State-owned mountain, coniferous forests, because it was converting them into wooded pastures, and later the prohibition was extended to cover similar forests owned by communes and private persons. These coniferous forests were divided into 8–25 cutting sections (generally 20) to be worked over in turn on a corresponding felling-cycle. In forests which had been cut over already felling was limited to two to three trees over a fixed exploitable dimension per acre, with some smaller trees where thinning was advisable. In the previously inaccessible forests the number was raised in some cases to as many as eight exploitable trees per acre in each felling-cycle. This regulation of the yield in selection forests by number of stems (*Possibilité par pieds d'arbres*) was continued in the Vosges until 1830 and in the Jura till 1840.

By 1760 all the forests of France had been brought under some form of regulated management, though as yet only on an empirical basis, and large areas had been restocked artificially by sowing, mainly with oak.

It was during the 18th century that the question of forest rights became acute in France. In the feudal period the tenants of agricultural plots had the right to graze cattle in the landowner's forest and also to coppice the outer portions of this forest for fuel. From the 14th century, when the central or high forest portions came under regulated (*tire et aire*) fellings, the survival of regrowth depended on the limitation of grazing. Many forest owners effected a settlement of their tenants' rights, either:

(1) buying up the tenants' rights in the high forest portion,
(2) setting aside part of the high forest for unrestricted grazing in exchange for the extinction of grazing rights over the rest,

or

(3) giving up the outer coppice areas to be communal forests, in exchange for all coppicing and grazing rights throughout the rest of the forests.

This third alternative was the one most usually adopted on the Continent. When the same position arose later in England the second alternative of giving up areas of high forest for communal grazing was generally adopted, the result being the commons of to-day.

From 1766 until the revolution of 1789 more harm was done to the forests. The population of France expanded to 26 millions, and on account of the shortage of wheat, Louis XV exempted from tax for fifteen years all newly-cultivated land. This led to the forest area of France being reduced by 1790 to one-sixth of the total area of France, the lowest figure in its history. In addition, during this period many high forests were converted to coppice with standards. Then during the revolution many people seized the chance to claim and exercise rights in a great many of the properties which had been confiscated from their owners by the Republic. Some of these rights were disproved later, but most of them were settled by the creation of more communal forests. The French State and private foresters did not believe that forests burdened with rights could be managed efficiently, and they preferred to give up considerable portions of them in order to obtain a right-free residue. When considering French forestry it must be remembered that State forests have always been in the minority. Proportions have varied, but the position to-day is that private individuals own about sixty-three per cent of the forest area, communes own about twenty-two per cent, and the State some fifteen per cent.

GERMANY

During these centuries there was no Germany as such, but there existed a number of separately ruled States and Principalities, each of which developed its own laws and traditions independently. Thus, reconstruction of the forests was not centrally controlled as in France, and some of the drawbacks of over-centralization were avoided. A practice similar to *sartage*, called *hackwald* in the German tongue, was developed in many areas, partly to augment local food supplies and partly to rejuvenate coppice. In the densely populated plains fuel areas were generally managed as simple coppice and the retention of standards, which were found to reduce coppice growth by their shade, was rare. There was more broad-leaved high forest on the hills than there was in France, and in these, concentrated felling in adjacent coupes was the common practice. In the coniferous forests of the mountains selection felling was tried in some areas, but more usually strips were clear-felled up and down the slopes in order to facilitate extraction. In the case of all concentrated fellings, sowings were carried out or, if seed was lacking, natural seedlings were transplanted from elsewhere. Also considerable afforestation of waste areas by sowing was undertaken.

However, the results were insufficient to keep pace with the demand for produce, and by the 18th century the need for regulated forestry had become so urgent that a series of ordi-

177

nances were issued laying down practices and enforcing the supervision of all forests by foresters of the local State or Principality. The earliest of these was the Hesse-Kassel Ordinance of 1711, which ordered that in broad-leaved high forest (which was principally of beech) seed and nurse trees should be retained at every ten to twelve paces, which means about forty to the acre. These trees were to be removed when satisfactory natural regeneration had been obtained. The good results of this practice attracted attention, and it was copied in other States, but not officially in France until the end of the century, though we have seen that some French foresters tried similar methods locally before that. In the Hesse-Nassau (Prussia) Ordinance of 1736 three successive regeneration fellings were prescribed for each coupe, a shade (seedling), a light (secondary), and a final felling. Local foresters had found that the most complete regeneration, particularly of beech, was obtained on a clean forest floor after a cautious thinning, that then ample light had to be allowed to penetrate the canopy to establish this regeneration, the growth of which, once it was established, was retarded unless the remaining trees of the old crop were removed. Thus, over two hundred years ago the principles of the Uniform or Shelterwood Compartment System were officially embodied in an ordinance.

However, in mountain coniferous forests the leaving of scattered seed-bearers standing in a felled area so often resulted in most of them being blown down, that, except in a few forests where selection felling was practised, there was a general return to clear-felling in strips. The artificial sowing adopted previously was, however, replaced by natural seeding from the adjoining unfelled forest. The prevailing winds generally blew up or down the valleys, so the strips could still be oriented up and down the slopes, which were approximately at right-angles to the valleys, in order to permit of extraction downhill. The direction of successive fellings was against the wind, so the newly exposed edges of the old woods were to the leeward and seeds from them were blown on to the felled strips.

The Uniform System was tried also in the Scots pine forests of the Rhine-Maine plains, but was not very successful. Either many of the seed trees were blown down and regeneration had to be completed artificially or, if many seed trees were left, their final felling damaged the regeneration, which un-like that of hardwoods, could not be cut back to reshoot. Accordingly, clear-felling in strips or in alternate strips at right-angles to the wind was prescribed.

The alternate strip system was also tried in other of the Prussian States, but in 1788 was abandoned in favour of the Uniform System, leaving a few seed-bearers only, to be felled after a fixed period of seven years, the first example of a definitely prescribed regeneration period. The Hesse-Kassel Ordinance of 1761 prescribed cleanings and laid down elabo-rate rules for thinning in high forests.

During the 18th century in Germany further large areas of waste land were afforested with Scots pine, Norway spruce, and some with oak. The tree seeds were sown with a crop of rye, or the area was planted after the rye had been reaped. This early form of *taungya*, called *Waldfelbau* (forest cultiva-tion), was advocated by von Langen in the Hartz mountains, and is also recorded in the Chiltern hills in England in 1745.

In 1765 a Thuringian forester called Oettelt demonstrated the desirability of thinnings from the economic aspect, by showing that early thinnings discounted against the cost of establishment at compound interest, had a great influence on the profitability of a crop.

It was also Oettelt who first expressed the capital, or Grow-ing Stock, of a forest mathematically in terms of the mature wood and of the rotation. He divided the volume of mature timber standing on a unit age-gradation area at rotation age by the number of years in the rotation, and thus obtained the mean annual increment of the final crop trees per unit. He called the volume of this mature timber V, but as it was only the volume of the final crop trees it is better that we should call it I and avoid the confusion which has been caused to foresters by the use of V for the measured volume of a number of different units. Thus he found that

$$\frac{I \text{ (vol. of final crop on a unit area)}}{R \text{ (no. of years in the rotation)}} = i \frac{\text{(m.a.i. of final crop}}{\text{trees) on a unit area}}$$

and, of course, $I = i \times R$.

He then assumed that i was the volume which each unit area of a complete series of age-gradations from one year to rotation age grew in each year, and that the volume of the growing stock of such a complete series, i.e. of a normal forest, could be expressed and summed in arithmetical progression,

$$i + i \times 2 + i \times 3 - \ldots\ldots + i \times R = (i + i \times R) \times \frac{R}{2},$$

or $\dfrac{(i \times R)\,R}{2} + \dfrac{i \times R}{2}$ at the end of the final growing season,

$\dfrac{(i \times R)\,R}{2} - \dfrac{i \times R}{2}$ just before the final growing season, and

$\dfrac{(i \times R)\,R}{2} = \dfrac{IR}{2}$ (or $\dfrac{VR}{2}$, as he called it) in the middle of the last growing season.

Although Oettelt called this the growing stock, it should be remembered that it represents only the volume of the eventual final crop trees standing at any one time on each of the age-gradations in a series, and then only if it were true that the final crop is in actual fact built up by equal annual instalments of increments throughout the life of a forest. It ignores all trees which are removed periodically as thinnings.

Oettelt also developed an earlier idea (that of Jacobi of Hesse-Nassau, in 1741) of making an allowance for variation in quality of locality. He reduced the area he cut annually in good sites and increased it in poor ones in order to cut equal annual yields.[1]

As he was working with oak and beech crops being regenerated by the uniform system he was unable to recognize each annual age gradation. He therefore distinguished between a periodic coupe and an annual coupe. He identified roughly on the ground six periods of the rotation, seedling, thicket, pole, small timber, medium timber, and mature timber. He then laid out on the ground equiproductive periodic coupes, using crops on the border lines between the various stages to adjust their areas. For the purpose of the annual felling he divided each periodic coupe into annual coupes. Thus was achieved the Framework (*Fachwerf*) Method, or allocation of woods to all periods of the rotation, which was a rigid method of management evolved by early foresters in an endeavour to bring order out of chaos.

[1] See page 99.

This brings us to the end of what may be termed the early periods of Western European forestry, passing from a stage of ample supplies till about the middle of the 14th century, through balance of supply and demand to the middle of the 16th century, of destruction and shortage to the middle of the 17th century, and of reconstruction on an empirical basis till near the end of the 18th century.

During this last period the British Isles lagged far behind. The earliest legislation of which we know is the Encoppicement Act of 1483, which legalized an earlier practice of enclosing coppiced areas against grazing. There were waves of replenishment activity in the New Forest and the Forest of Dean, generally after wars, notably about 1600 and 1660. The offers of medals and premia for planting by the Royal Society of Arts in 1758 stimulated a good deal of private planting.

CHAPTER

20

LATER PERIODS. THE RISE AND INFLUENCE OF FOREST
SCHOOLS

The development of communications by canal and road reduced reliance on local supplies of forest produce. Coal began
to replace wood and charcoal as fuel and later came the development of railways and of overseas trade. Coal and other
mining created a demand for pitwood, and the spread of
reading a demand for pulpwood. The lure of these markets
led to the domination of silviculture by economics.

The earliest forest schools were founded at the end of the
18th century by individuals and later were taken over as the
official State schools of their countries.

The Prussian School[1] In 1789 Georg Hartig (b. 1764, d.
1837) founded a school in his home town of Hessen, in
Wurttemberg, and moved it to Stuttgart in 1807. In 1811 he
was appointed head of the Prussian Forest Service and took
his school with him to Berlin, and later to Eberswald, its
present site.

Hartig was the great advocate in Germany of the Uniform
System, which he introduced throughout Prussia. He was dealing mainly with oak and beech forests, but also with Scots
pine stands, the results of previous afforestations, which he
greatly extended. He developed the idea of dividing a forest
into convenient Felling Series, each containing woods repre-
sentative of all age-classes. Series by series he allocated the
stands to all periods of the rotation as Oettelt had done, but
by volume. He did this by enumerating all stands and allocating them provisionally to periodic coupes or blocks in a
Trial Working Scheme. Then he added the mean annual increment which he anticipated the final crop trees in each
block would put on up to the time of felling shown in the trial

[1] There were earlier schools which did not last, e.g. that started by
Zanthier in 1763.

182

scheme. For example, with four twenty-year periods, ten years' increment would be added to the volume of Periodic Block I because that would be the average number of years the trees in it would stand, thirty years' increment to P.B. II, and so on. The total of the growing stock and increment so obtained was then divided by the number of regeneration periods in the rotation, and each block adjusted to yield that amount at maturity. This reallocation of stands to blocks, which, of course, involved recalculation of the increment for any stand moved from its place in the trial scheme, was called the Final Working Scheme. The annual yield in regeneration fellings in P.B. I was then the volume of that block plus its increment for half the regeneration period, divided by the number of years in the period. Hartig also drew up careful silvicultural rules for thinnings in the other blocks, estimating the probable increment other than final crop increment, and emphasized that the thinning yields were additional to the final yield.

The method was cumbersome and tedious to apply. It relied on calculations of increment made for long periods; for instance, in the case of the youngest block for the whole rotation less half a regeneration period, which could not possibly be reliable.

As a result of the earlier fellings in adjacent coupes, the crops due for regeneration were generally contiguous, and Hartig aimed at perpetuating this, though it frequently involved felling some stands which were immature and leaving others to grow on longer than they should, just to make up orderly blocks. With oak and beech seeding freely every nine to twelve years he fixed twenty years as the regeneration period and divided his rotations into twenty-year periods.

In 1836 Hartig's successor, von Reuss, abandoned the allocation of woods to all periods of the rotation, applied an age-class check, and drew up plans for the next regeneration period only, ordering revision at short intervals. Later, Judeich's ideas of selecting unthrifty crops as well as mature ones for regeneration was copied. Thus, only the silvicultural side of Hartig's work survived in Prussia, though, as we shall see later, in France his teachings were developed to their logical conclusion, with one important modification, the substitution of area for volume as the basis of periodic blocks.

The Saxon School   In 1795[1] Heinrich Cotta (b. 1763, d. 1844) started a private school at Zillbach, in Thuringia. In 1810 he was appointed head of a forest surveys school in Tharandt and moved his own school there. In 1816 it became the Royal Forest School of Saxony.

In 1796 a Bavarian forester named Schilcher, working in hilly districts, drew attention to the fact that accidents, particularly windfalls among seedbearers, might upset long-range plans. He proposed to leave closed forest between successive regeneration coupes, only returning to fell adjacent to and to the windward of a previous coupe after completion of its regeneration and the removal of its seed trees. This involved the division of the regeneration block into cutting sections.

Cotta developed and elaborated this idea in Saxony in 1811. He divided each forest by numerous parallel rides N.–S. or N.W.–S.E., and by less numerous rides at right-angles into rectangular compartments. The rides facilitated not only extraction, but also the orderly grouping of compartments into cutting sections. He grouped all the compartments under regeneration into as many cutting sections as there were years in his felling-cycle. Thus, if four successive fellings were needed to complete regeneration in a twenty-year period, he fixed the felling-cycle at five years and divided the regeneration block into five cutting sections which he worked over in turn against the wind. He had already abandoned Oettelt's division of his periodic coupe, or regeneration block, into annual coupes.

Cotta sought to avoid the difficulties of Hartig's volumetric method by dividing his forest into periodic blocks by area instead of by volume, and in 1820 devised his so-called Simplified Method. He studied the quality of the localities and endeavoured to allocate equiproductive areas to each period of the rotation. He selected a regeneration period sufficient to establish regeneration by successive fellings, and which at the same time was a sub-multiple of the rotation or exploitable age. He advocated thinnings and tendings on purely silvicultural grounds in the blocks not undergoing regeneration,

---

[1] Some writers say 1785. As he was only twenty-two years of age then, the later date, quoted by others, has been adopted. Fernow, in his *History of Forestry*, gives both dates on different pages.

and fixed for them a cycle which was a sub-multiple of the regeneration period. He did not visualize that the allocation of compartments to periodic blocks should be permanent.

At the beginning of each period he measured the standing volume of all the woods to be regenerated in the period, i.e. of those compartments allocated to P.B. I and, recognizing that some of the trees would be cut at once and that some would stand over till the end of the period, he decided the simplest way of estimating the increment which would be harvested was to allow for increment on all the trees for half the period, the average time for which they would stand. Thus, he fixed his annual yield as

$$\text{A.Y.} = \frac{\text{Vol. of P.B. I at start} + \text{est. increment on this Vol. for } \frac{1}{2} \text{ Period}}{\text{Number of Years in the Period}}$$

This, of course, was only the final yield from regeneration fellings. The volume felled annually in thinnings from the other periodic blocks was additional to this and was estimated at the average of such thinnings in past years.

This method of calculating the final yield has persisted under the name of Cotta's method. It was the same as Hartig used, and it is sometimes said that Hartig copied Cotta, but the reverse is probably true.

In poorly-stocked hardwood forests suffering from past mismanagement Cotta abandoned the Uniform System and began to clear-fell the regeneration blocks and replant them with Norway spruce on economic grounds. As usual he divided the blocks into cutting sections and felled against the wind, planting four to five year old nursery transplants among the hardwood stumps. By 1817 he had abandoned in these hardwood areas undergoing conversion, the allocation of stands to all periods of the old hardwood rotation, because this was liable to delay the conversion of poor woods into economic crops. He adhered strictly to the principle of cutting sections, but concentrated more on what should be felled and converted in the immediate future than on the long-term plan. As the rotation of the spruce he was planting would be shorter than that of the hardwoods he was replacing this appeared reasonable. Accordingly he advocated short-term felling plans with frequent revisions and the maintenance of control records. When applied to forests being worked under the Uniform System this development involved a distinction between the working plan revision period and the regeneration period.

In 1833 von Klipstein, of Saxony, became nervous lest foresters, while concentrating rightly on the fellings of the immediate future, should lose sight of the long-term plan and create abnormal age-class distribution. He insisted that in all forests a comparison of the actual and the normal age-class distribution should be made before it was decided what stands were to be regenerated or converted in the next working plan period. Yield tables had been devised more or less simultaneously in Prussia (by Kregting) and in Austria (by Trunk) and developed by all German foresters, so using these he followed Schilcher's idea of 1796 and reduced all the areas of a forest to terms of one quality class. The total reduced area of the forest divided by the number of age-classes adopted gave the area of a normal age-class or periodic block. A comparison of this normal age-class distribution with the actual, similarly reduced to terms of the same quality, indicated whether and in what way the forest was understocked or overstocked. There was then a sound basis for decisions to be made about the rational reinvestment of the forest capital. It was recorded above that von Reuss adopted the same idea in Prussia in 1836, whether independently or not is not known.

In 1860 Judeich, Professor at Tharandt, 1866–94, drew attention to the frequency and the serious results of accidents caused by wind, snow, etc., in the forests. He developed stand management (*Bestandwirtschaft*) based on the conditions of individual crops. He selected and listed the stands which, irrespective of age, he thought should, if possible, be felled and replaced in the ensuing regeneration period. He then applied an age-class check and compared, not only the normal and actual age-class distribution at the beginning of the period, but also what they would be at the end of the period, if fellings were carried out in accordance with his provisional list. If this check indicated that some of his proposed fellings would cause serious fluctuations in the future periodic yields expected in the long-term plan, he revised the list and rechecked it until a satisfactory compromise between the short-term plan and the long-term one was reached.

Towards the end of his life Cotta became nervous about the extension of pure spruce planting on the plains and foothills of Saxony where, outside its natural mountain habitat, it was particularly susceptible to wind, drought, and insect attack. His warnings were not heeded, however, and under the influence of Pressler, who was the great advocate of maximum net money returns and the soil rent theory, conversion of mixed high forest to pure spruce for the production of pitwood and pulpwood on short rotations continued.

The practice was copied in Bavaria, Bohemia, Austria, and the German-speaking cantons of Switzerland, and led to serious silvicultural and financial difficulties later on.

The great contribution of Cotta's school was the advance from allocation of woods to all periods of the rotation to allocation only to a short working plan period, with an age-class check and a review of the probable effects of the immediate proposals on the long-term plan. He introduced elasticity by demanding that no uneconomic crop should be retained longer than was necessary to avoid violent fluctuations in future yields. By concentrating on the planting of pure spruce for immediate financial reasons the school caused economic difficulties in the future, even though Cotta himself saw the red light too late.

These two forest schools attracted students from all over Europe, but there is no evidence that they had any connection with the foundation of the Russian Forest Institute, opened in St. Petersburg in 1803. This Eastern European forest school appears, however, to have had no influence on Western European forestry, and unfortunately little is known about its teachings at that time.

The French School In 1824 the French State Forest School was founded at Nancy, under an Alsatian, Bernard Lorentz (b. 1775, d. 1865), who had studied under Hartig in Prussia. In 1825 Parade (b. 1802, d. 1864), a pupil of Cotta, was appointed a lecturer.

After the revolution there was a great deal of forestry activity in France. Reclamation of the swamps of the Landes district, in the south-west, was commenced by sowing Maritime pine, and the sowing of silver fir in ruined mountain forests, and of oak elsewhere, with or without field crops, again became widespread. The number of seedbearers to be retained in *tire et aire* coupes in beech and oak high forest was increased to thirty to forty per acre, as in Prussia. To Lorentz, however, when he first came to Nancy, the most vital question was the future of the huge areas of coppice with standards. He believed fervently that this system was 'essentially a vice', and that coppice growing and timber growing should be carried out on separate areas. Further, he saw that the availability of coal was already beginning to affect the market for wood fuel, and he thought that short rotation fuel growing should be left to communal and private forest owners, while the duty of the State was to grow timber on long rotations in its forests. In the areas he wanted to convert to high forest the coppice was from one to forty years old and good seedling regeneration could not be obtained from trees younger than sixty to seventy years. He therefore advocated that conversion should start by stopping all coppice cutting and allowing the coppice to grow up into the canopy with the standards for forty years, during which there should be two light thinnings only. After this he intended to regenerate the forests by the uniform system in 100 years. By 1839 Lorentz had succeeded in having some 123,000 acres of coppice with standards in State forests put into reserve for conversion. This was unpopular with the Treasury because of the drop in forest revenue and also with the masters of forges, salt works, etc., who saw the price of firewood, of which over 282,000,000 stacked cubic feet was still being used a year, would rise. The opposition won and Lorentz was forced to retire in 1839.[1] He was followed at Nancy by Parade, who agreed with his views in general but was rather more tactful, and in 1843 brought out his Method of Direct Conversion (now sometimes called the Classical Method). In this the conversion period was to be equal to the future high forest rotation, generally 120 years, and was divided into four equal periods, the forest being divided into four periodic blocks of equal area. Three of these blocks continued to be worked on the coppice rotation with the reservation of

[1] He was then head of the forest administration in Paris.

an increasing number of standards in order to reduce the vigour of the coppice, and so its eventual competition with the future seedling regeneration. In the other block, the first conversion block, the coppice was cut to obtain natural regeneration from the standards, planting was resorted to where seedlings failed to appear, and suckers which interfered with seedlings and plants were grubbed up. The standards were to be cut gradually as regeneration advanced. A second block was to be converted in the second period and so on. The drop in revenue was smaller, so the Treasury did not object: the supply of firewood was less affected, and as the use of coal was becoming more general the industrialists did not oppose the idea. The story has been told to indicate that foresters cannot afford to neglect economic and political aspects if they wish to continue to practise.

Both Lorentz and Parade, having been trained in German schools, believed in the uniform system of management for high forests and strove to introduce it into France. Parade introduced it in 1830 under the name of the Method of Natural Regeneration and Thinning, an appellation which did not survive, and by reason of its early promise it was officially sanctioned for forest after forest from 1840 onwards. Parade taught the division of each forest into felling series, and in each series the allocation of homogenous compartments to periodic blocks. These blocks were not necessarily equiproductive in existing growing stock, but were of equal surface area, excepting that, when there were considerable differences of site likely to affect production permanently, he made the blocks containing the poorer sites larger. They were self-contained and contiguous blocks based on the old *tire et aire* coupes and their selection inevitably involved some sacrifice of potential increment here and there in the interests of convenience and order. The regeneration period was fixed at a fraction of the rotation, generally at ten years in the western regions, where oak seeded abundantly every five to six years, and at twenty years elsewhere. The annual yield in the regeneration block was calculated according to Cotta, but the choice of compartments within the block in which it was harvested each year by the various regeneration fellings was left to the executive officer.

Thinnings in the other blocks were carried out on a cycle

or cycles chosen on silvicultural grounds, and were purely silvicultural.

By the middle of the 19th century the use of the system was compulsory in all high forests in France, hardwood and coniferous, excepting only those of Scots pine and Maritime pine. Later modifications were the variation of the regeneration period forest by forest according to the frequency of mast years and general local condition (1859), and the definite presumption that the periodic blocks should be permanent allocations of compartments for all time, to be felled and regenerated in a cycle of regeneration periods. The names Method of Even-aged High Forest and Method of Permanent Periodic Blocks were adopted for the system. Thus, the French love of order developed Hartig's original ideas about management to their logical conclusion.

In 1864 a particularly bad storm in the Vosges mountains, which razed to the ground in one night the yield prescribed for several years, brought to a head earlier objections that regeneration in the mountains would not confine itself to the regeneration blocks only. The result was a decision that the uniform system should be continued, but in the Vosges with a different method of yield control, based on the enumeration of all measurable trees all over the forest, in order to permit the inclusion of windfalls outside the regeneration block in the prescribed yield.

Before 1858 a working plans officer called Masson had argued that if V was the enumerated volume of the growing stock found in a selection forest, of which the average age ought to be half the rotation or exploitable age R, the increment laid on each year was $\dfrac{V}{\frac{1}{2}R}$ $\left(= \dfrac{2V}{R}\right)$ and this amount should be prescribed to be cut annually.

This expression was the same as that based on Oettelt's theory, but in this case the growing stock was measured, instead of being calculated from the measurement of the rotation-aged age gradation unit by assuming that the volume of this unit was equal to the sum of the increments laid on to each age gradation unit annually. It assumes, however, that all age-classes are present in the normal proportions and therefore that the average age is half the rotation, which is unlikely to be the case. The question of whether the normal age-class distribution in a selection forest is the same as in an even-aged forest need not be considered here, because the regulation was to be applied to uniform forests. Masson considered that if there were too many old trees in a wood the annual increment would be less than $\dfrac{V}{\frac{1}{2}R}$ and that cutting this amount annually would reduce the overstocking. Conversely if the proportion of younger trees was excessive the annual increment would be higher than the cut and the stocking of the forest would be increased.

In other mountain forests of France, including those of the Jura, the ordinary uniform method was applied for another thirty years. In fact, there is one Jura forest, La Fuvelle, which is still managed with the permanent periodic blocks laid out by Broillard in 1858, though there was one ten-year period during which the yield was regulated by volume instead of by area and volume. Broillard was an Inspector of Forests who had advocated the permanence of blocks for some time before he became a lecturer at Nancy and published his idea in 1878.

South Germany The need, particularly in mountain forests, to base the annual cut on the actual volume of the growing stock and its increment was realized in Austria in 1788. A distinction was made between the normal and the actual growing stock, and Oettelt's formula, G.S. $= \dfrac{I \times R}{2}$, was used to calculate the former, while the latter was measured. The famous, but anonymous, official of the inland revenue proposed that the correct annual yield from a forest was,

$$\text{Normal Increment} - \frac{\text{Normal Growing Stock} - \text{Actual Growing Stock}}{\text{Number of Years in the Rotation}}$$

when the forest was understocked, or

$$\text{Normal Increment} + \frac{\text{Actual Growing Stock} - \text{Normal Growing Stock}}{\text{Number of Years in the Rotation}}$$

when the forest was overstocked.

The intention was that the forest should in each case be brought to a normal volume in a rotation by cutting more or less of the increment. However, it took no account of the age-class distribution of the stock and so might fix a yield which could not be cut in mature trees. It also assumed that normal increment, i.e. the volume of the rotation-age age gradation, was being put on each year.

In some of the mixed broad-leaved and coniferous forests of the hilly districts of Wurttemberg, clear-felling in strips up and down the slopes against the wind favoured the conifers, especially spruce, at the expense of the heavier-seeded oak and beech. The uniform system practised in some of these forests favoured the shade-bearing species, particularly silver fir and beech. A compromise, called the Shelterwood Strip System, was developed, and from 1828 became standard practice in Wurttemberg. Each forest was divided into felling series and in each series compartments were allocated to a regeneration block. This block was divided into as many cutting sections as there were years in the cutting-cycle between successive shelterwood regeneration fellings. These fellings were carried out in narrow strips against the wind. Stands were not allocated to all periods of the rotation, and there was no age-class check, but reliance was placed upon a theoretical formula, about which we need not trouble, to bring the capital to normal. A sound feature was that windfalls of mature trees outside the regeneration block were counted against the prescribed yields, while silvicultural thinnings in such areas were not.

The need to take notice of age-class distribution was recognized later, and in 1862 Judeich's individual stand system with a modification by Hufnagl was introduced in Wurttemberg, as well as in Hessen and Prussia. In this a list of all woods which should, if possible, be regenerated during a definite period of the rotation, because they were mature or unthrifty was drawn up. The average age of the forest was then ascertained by finding the area and average age of each stand, multiplying these two figures together, summing the results, and dividing the sum by the total area.[1] The size of the block which should be regenerated in the period proposed, if the

[1] See p. 110.

age-class distribution were normal, would be the area of the series divided by the number of such periods in the rotation. This size was modified by the relationship the actual average age of the series bore to the normal average age, which, of course, should be half the number of years in the rotation, i.e. Area of Regen. Block =

$$\frac{\text{Area of Felling Series}}{\text{No. of Periods in the Rotation}}$$

$$\times \frac{\text{Actual Average Age of Growing Stock}}{\text{Normal Average Age (i.e. } \frac{1}{2} \text{ Rotation)}}$$

The provisional list of stands to be regenerated was then adjusted to this size, which would be big if there were an excess of old woods and small if most of the stock were young.

Clear-felling was prohibited in the higher parts of the Black Forest in 1883, because of erosion and windfall. In these areas the selection system was applied and the annual yield was regulated by a formula devised by von Mantel which was the same as that which Masson had worked out for the Vosges forests, i.e. that the annual yield should be equal to the measured volume divided by half the rotation. It is said, however, that the yield so calculated regulated the felling of mature trees only and that all younger trees cut in silvicultural thinnings were additional to it. If so the felling would be less conservative than that of Masson, and as we shall see later, nearer to the actual possibility of the forests.

On the lower slopes of the mountains the shelterwood strip system by cutting sections was used, and stands were allocated by area for regeneration in the ensuing period only. From 1869 the yield was regulated by a modification of the Austrian formula devised by Heyer, of Bavaria, some years earlier. Heyer substituted the actual mean annual increment for the normal increment and calculated his annual yield for a period of 'a' years, during which he thought the growing stock could be brought to normal, but revised it at frequent intervals. His annual yield was

$$\frac{\text{Actual G.S.} + \text{M.A.I. of all stands for } a \text{ years} - \text{Normal G.S.}}{a \text{ years}}$$

The actual growing stock was measured, the annual increment of each stand was its volume divided by its age, and the normal growing stock was the volume of the rotation and age gradation multiplied by half the rotation. It was, therefore, still fairly empirical.

Since the increment added was that for all stands, whether for regeneration or thinning during the period, the yield was a total one covering both regeneration fellings and thinnings.

Bavarian foresters recognized that frequent windfalls were inevitable in the old crops to the windward of their shelter-wood strips, and that these often resulted in advance regeneration of the shade-tolerant species, silver fir and beech. They realized that these windfalls must be counted as part of the prescribed yield, and that the advance regeneration they caused brought a desirable mixture of shade-bearers into the new crops, which otherwise tended to consist almost entirely of the light-demanding spruce. They found that the presence of a proportion of silver fir and beech increased the wind resistance to the crops.

This brings us to the end of what can be considered as another period closing round about 1870, after which the current practices evolved.

CHAPTER

21

Experiences such as the great storm of 1864 in the Vosges made a number of French foresters doubt the suitability of the uniform system for the management of mountain forests. A Jura forester named Gournaud felt so strongly that his forests should be managed by selection, that he resigned from the State service in 1866 because he was not allowed to adopt this method. More will be heard about his influence later. However, towards the end of the first regeneration period of several of the working plans a change to selection was permitted in several mountain forests.

To regulate the yield in these cases a new method of calculation[1] was set out in an administrative note dated 1883, which was anonymous but is generally attributed to Mélard, who as Inspector-General of Working Plans expanded it in a signed note of 1894.

The method ordered was the determination of an exploitable age, and from this a corresponding exploitable diameter, followed by enumeration down to trees of one-third of this diameter. The trees enumerated were classified into a Large Wood group, two-thirds of exploitable diameter and up, and a Medium Wood group, between one-third and two-thirds of exploitable diameter. The theoretical proportions of the enumerated growing stock falling into these two groups, assuming that they corresponded with the thirds of the rotation, and assuming an even mean annual increment throughout the life of a stand, were five parts by volume for the large group to three parts for the medium group. This proportion is not far out if a volume curve based on current annual incre-

---

[1] It is probable that the idea of division into size-groups was of considerably earlier date.

ment is substituted for the straight line based on m.a.i. (Fig. 7).
It does, however, vary according to species, site, and rotation.
Light demanders such as Scots pine growing on good sites
reach more than a third of the rotation diameter in a third of
a rotation, and more than two-thirds of the diameter in two-
thirds of a rotation, but do not do so on poor sites. Shade-
bearers such as silver fir grow slowly at first on all sites, but
after the first third of a rotation they grow faster and reach
two-thirds of rotation diameter in less than two-thirds of a
long rotation, but not of a short one. Spruce on medium sites
increases in diameter fairly constantly with age throughout
most rotations. These facts suggest that the method could be
made more accurate by substituting the actual average size
limits corresponding with the respective thirds of rotations
in each forest for the assumed sizes. As the method is used
mainly for mixed forests of spruce and silver fir it might be
necessary to differentiate between the species. Presumably
the presence of silver fir tends to cause the volume measured
as the Large Wood group to be more than that grown in a
third of the rotation, and so suggests over-stocking. There is
also the question of the difference between size-class distribution in all-aged forest and in uniform forests. As space in an
all-aged forest is shared by trees of different sizes such a forest
can carry a larger proportion of its growing stock in large
trees than can a uniform forest. When judged by the test of the
5 : 3 ratio an all-aged forest would therefore appear to be
over-stocked with large trees. Certainly many of the French
silver fir and spruce forests have been considered to be over-
stocked.

French foresters enumerated some 100,000 acres of forests
which they considered satisfactorily stocked and found approximately the 5 : 3 proportions. It was, therefore, argued
that the large wood group could be felled and replaced in a
third of the rotation, during which time the medium group
would become the next large group. When the method was
first used and the proportions of 5 : 3 were not found to exist
between the groups transfers of size-classes were effected to
make them so. Often, indeed, the enumerated growing stock
was grouped in the desired proportions for the purpose of
yield calculation. The annual yield calculated was therefore

$$\frac{\frac{5}{8}\ \text{Measured Volume}}{\frac{1}{3}\ \text{Rotation}} \text{ which equals } \frac{1\cdot 875\ \text{Measured Volume}}{\text{Rotation}},$$

actually less than Masson's $\dfrac{2V}{R}$.

All enumerated trees either cut or blown down were counted against this yield, and before long it was realized that it was too conservative. By 1894 the increment of the large wood group for a sixth of the rotation, the average time a tree would stand, was added. The large group was aged from two-thirds to full rotation age and theoretically its age would be $\frac{1}{2}\ (R + \frac{2}{3}R) = \frac{5}{6}R$. As it had grown to its present volume of $\frac{5}{8}V$ in $\frac{5}{6}R$ years its m.a.i. was $\dfrac{\frac{5}{8}V}{\frac{5}{6}R}$ and assuming that it grew at this rate for the $\dfrac{R}{6}$ years it would stand, the total amount of timber available for cutting in a third of the rotation expressed as an annual yield would be

$$\frac{5}{8}V + \frac{\left(\dfrac{\frac{5}{8}V}{\frac{5}{6}R} \times \dfrac{R}{6}\right)}{\frac{1}{3}R} = \frac{\frac{5}{8}V + \frac{1}{8}V}{\frac{1}{3}R} = \frac{\frac{6}{8}V \times 3}{R} = \frac{2\cdot 25V}{R}$$

The 1894 note suggested that normally no account would be taken in the yield calculation of any increment put on by the medium wood group, some of which would undoubtedly have to be realized as thinnings. The intention was to keep on the safe side and that if there were an increase in capital due to cutting too conservative a yield this could be corrected if desired when the plan was revised. Later it was found that the yields prescribed were generally too conservative and a working plan officer was authorized at his discretion to include a portion of the medium wood group increment, usually one-third if the forest was considered to be normally stocked. By then increment rates used in the calculation had been more or less standardized at one per cent for the large group and three per cent for the medium group, so using the figures per acre (true cu. ft.) which had been accepted for normal stocking in the Vosges the annual yield would be calculated for a 150-year rotation as follows.

A.Y. from present L.W. cut in a third of $R =$

$$\frac{2{,}860}{50} = 57 \cdot 2 \, \text{cu. ft.}$$

A.Y. from half annual increment of L.W.

$$\frac{2{,}860 \times \cdot 001}{2} = 14 \cdot 3 \quad \text{,,}$$

A.Y. from one-third annual increment of M.W.

$$\frac{1{,}716 \times 0 \cdot 03}{3} = 17 \cdot 2 \quad \text{,,}$$

Total Annual Yield $\qquad\qquad\qquad = 88 \cdot 7 \, \text{cu. ft.}$
which is $1 \cdot 94$ per cent per annum of the growing stock
(4,576 cu. ft. per acre) and equal to $\dfrac{2 \cdot 9 V}{R}$.

Still later, in the 1920s, increment figures based on actual studies in each particular forest began to be used in the calculations. Thus, we find increment percentages of from $1 \cdot 2$ to $1 \cdot 5$ per cent for the large wood groups, and of $3 \cdot 3$ to 4 per cent for the medium groups. The conservative yields fixed in the latter part of the last century sometimes lead to an increased stock of large trees which has either been accepted by a lengthening of the rotation and an increase in the exploitable diameter, or reduced by special measures. One such special measure is the formation of an extra group of trees over the exploitable diameter, the Very Old Wood group, to be realized in a period of twenty years plus its increment (generally at half per cent) for ten years, in addition to the annual yield calculated for the other two groups. In forests for which a selection system was sanctioned a felling-cycle was fixed on silvicultural grounds at any fraction of one-third of exploitable age, and compartments were worked over each year in accordance with a list drawn up, until the prescribed annual yield had been obtained. This meant that the felling-cycle could not be adhered to strictly, that it was lengthened when there was more mature timber to cut, and shortened when there was less. Later a normal curve of the desired distribution of size-classes, obtained from enumerating an area of the forest which appeared to contain all the size-classes in reasonable proportions, was provided so that the marking officer could compare it with the curve of actual size distribution in each compartment, and mark accordingly to favour the deficient classes and reduce the others.

By 1883 it was realized that the retention of permanent periodic blocks in even-aged forests involved too many sacrifices and that it was generally better at the end of a period to reallocate compartments to blocks according to their maturity at the time. This was called the uniform system with revocable blocks, but really it was what both Cotta and Parade had intended all along.

In many of the spruce, silver fir, and beech forests of the mountains for which the selection system was not authorized, this modification was insufficient and in 1894 Mélard abandoned in them both the idea of allocating stands to all periods of the rotation, and that of regenerating a particular area in a particular time.[1] Enumerating the whole growing stock of a felling series and grouping the size-classes he calculated by his famous formula the total annual yield of regeneration fellings and thinnings for a third of the rotation. This yield was accepted for a working plan period of not more than twenty years and not less than ten years. He divided the compartments, sometimes even sub-compartments, into two groups,

- (a) those in which regeneration should be carried on or started during the working plan period: the regeneration area or group, coloured blue on the map (hence the name *Quartier Bleu* given to the system).
- (b) the rest to be tended or thinned during the period: the thinning or tending area or group (*Quartier Blanc*), left white on the map.

He indicated the order of regeneration fellings in (a), which order could be varied as seed production and progress of regeneration indicated from time to time.

He fixed a thinning and tending-cycle, drew up a table of areas in (b) to be treated annually on this cycle, and estimated the annual volume to be cut in thinnings.

He deducted from the prescribed annual yield each year the measured volume of all accidental falls of enumerated

[1] See pp. 125–128.

trees and of all thinnings made, and used the balance of the yield for regeneration fellings.

This division of the yield tended to increase the proportion of large trees which was at first thought desirable, but later it was found that regeneration was slowed down unduly, and that pure silver fir was tending to appear instead of the light-demanding spruce, which was more valuable. In 1912 a conservator in the Jura named Schlumberger advocated a different allocation of the total yield. He thought that one-half to two-thirds of the prescribed yield should be allocated to the regeneration area, and the balance to the thinning area, depending on the state of the forest. Accidental falls were to be measured and debited first against the allocation of the area in which they were found, before the balances were used in regeneration fellings and thinnings respectively. This meant that a definite area could not be thinned each year, and in some forests thinnings fell badly into arrears. More recently it has become usual to estimate what windfalls are likely to occur in an average year in each area, how much thinning out to be done to maintain a reasonable thinning-cycle, and then let the balance be used for regeneration fellings.

Another refinement introduced has been the division of the thinning area into two, one part to be those compartments which will be due for regeneration next, where thinnings will be in the nature of preparatory fellings and can be delayed if necessary (*Quartier Jaune*). In these forests, which were very subject to wind damage, it was at one time customary to deduct from the enumerated volume of the large wood group before calculating the yield a 'technical reserve', to which all windfalls over an average figure allowed for in the allocation of the yield could be debited (or to which the amount by which windfalls fell short of the estimated figures in any year could be credited). The idea of this was that when bad storms damaged one part of a forest, thinnings and regeneration fellings in the rest need not be held up. It should not be confused with the reserve of one-quarter of the yield which has to be left uncut in all communal forests by law, until the commune obtains sanction to cut its accumulation for some special local public work.

An interesting example of the use of a combination of

methods of yield regulation is seen in the 1938 revision of the plan for the only Jura spruce and silver fir forest (La Fuvelle) in which permanent periodic blocks are maintained. Seven of these were laid out by Broillard in 1858 for a 140-year rotation, and the revision was for the fifth period. The whole growing stock of the forest was enumerated and a total yield (147 cu. ft. per acre per annum = 2·3 per cent of G.S.) was calculated according to Mélard. A final yield, according to Cotta, was then calculated from the volume standing on P.B.V. and the difference between this and the total yield was to be harvested in thinnings from the other blocks on a ten-year cycle. The regeneration yield came to forty-seven per cent of the total. As the regeneration period of twenty years is rather short for the site, there is a special prescription that during the first five years of the period the stock on the whole of P.B.V. was to be reduced from the 8,200 cu. ft. per acre it was carrying to 5,400 cu. ft. per acre, which was considered to be a suitable density for seedling establishment. From these figures the average of regeneration fellings for each of the first five years was calculated, and by deduction from the prescribed yield from the block, the balance left for each of the other fifteen years.

The uniform system is said to be responsible for the production of 10 per cent of the total out-turn from this forest in clear wood with annual rings not exceeding 2 mm. wide, which sells for special-purpose plywood at seven to eight times the price for ordinary saw timber, of eighty per cent in saw timber and ten per cent in pulpwood. The belief that this could not be done under the selection system explains the reluctance of French foresters to give up the uniform system.

In the hardwood forests also the uniform system with permanent or even with revocable blocks proved too rigid and regeneration fell into arrears, except in the western areas, where oak mast years were frequent. In 1890 Puton[1] proposed allocation of compartments to one period only, the single periodic block method. In 1900 Duchaufour developed the

[1] Puton wrote in 1890 'to tell the truth a single periodic block is sufficient, the first dedicated to regeneration felling. The others will not be needed for thirty, sixty, or ninety years. What good does it do then to determine them in advance?'

method. He first calculated the normal area of a periodic block as the total area of a series divided by the number of regeneration periods in a rotation. He then picked out from all over the series the compartments and sub-compartments in which regeneration had been started but not completed, and reduced their area by the ratio which the basal area of growing stock on them bore to the basal area which the same surface area would carry if normally stocked. Deducting this reduced area of the partially regenerated compartments from the normal area fixed for the periodic block gave the area of mature, untouched stands which could be added for regeneration during the period. If the area of such mature stands was insufficient to make up the full normal area of the block, the plan provided for the regeneration of a block smaller than normal, that is to say, no immature woods were ever prescribed for regeneration. The final yield for the period was calculated according to Cotta from the measured volume of timber in the periodic block plus its increment for half the period. The rest of the felling series was subjected to silvicultural thinnings on an area basis, determined by the length of the thinning-cycle selected, the thinning yield being additional to the final yield. In revisions of working plans for even-aged oak and beech forests this single periodic block method was applied widely from 1913 onwards. The scattered compartments included in the regeneration block were generally painted blue on the map, and were sometimes loosely referred to as the *Quartier Bleu*. It would have been more accurate to call them the *Affectation Bleu*, the blue block, instead of the blue area, and the use of the latter term has sometimes lead to confusion with Mélard's regeneration area method. Duchaufour's method is one of regenerating a definite acreage in a definite number of years, with the final yield controlled by the area of a periodic block and the volume standing on it. Mélard's 1894 method consists of indicating those compartments in which regeneration operations will be carried on during a short working plan period, without any definite commitment as to the area they will cover, and with a total yield calculated from the volume standing on the whole felling series and the rates of increment of two size-groups. In both these methods frequent revisions of the plan and resort to planting, when regeneration is delayed, are now the rule. In their practical execution the difference between them is not great.

It is now necessary to look back to the progress of conversion of coppice with standards to high forest, the starting of which about 1830 had caused such a furore.

The method of direct conversion which Parade started in 1843 failed, especially in the north-eastern regions, where oak mast years were infrequent. Much planting had had to be done in the first conversion blocks, coppice and sucker growth had tended to smoother seedlings and plants, and many cleanings had been necessary. It was realized that conversion methods would have to be adapted to local conditions, and several different systems were evolved for different parts of the country.

Method of Conversion with Waiting Period   From 1873 onwards the first conversion plans reached the end of their first thirty year periods and fell due for revision. Instead of carrying on with the conversion of the second blocks these were subjected to a waiting period for the next thirty years, during which light thinnings only were carried out with the aim of freeing all standards. In the other blocks the usual coppice cuttings were carried on, but with the reservation of all possible standards. In the next period the block which had been subject to a waiting period was to be regenerated by shelterwood fellings, and it was anticipated that the coppice would not be so troublesome by then. In new areas taken up for conversion a future high forest rotation of 160 years was often adopted, and the forest was divided into four blocks, the first, to be given a waiting period of forty years, consisting of the compartments with the greatest number of standards.

Where nothing better can be done this method is still practised, despite the drop in immediate revenue which it involves.

Method of Conversion with Intensive Reservation   In Normandy, where oak seeds well every five or six years, it is possible to reserve many seedlings in the ordinary coppice coupes. From 1880 Aubert introduced a method whereby old standards were felled and intensive reservations of maiden trees were made in each annual coupe as its turn for coppicing arrived. He organized thinnings, selected the best resulting crops to grow on for a hundred years or more, and regenerated the poorer crops by the shelterwood system when they were sixty years or so old. This causes little drop in revenue and is popular wherever a reasonable number of seedlings are to be found in coppice coupes. Periodic blocks are not formed till after conversions.

Method of Conversion with Introduction of Conifers   On the lower slopes of mountains it was often found that hardwood coppice with standards was invaded by coniferous seedlings when there were any such seedbearers in the vicinity. Therefore in suitable forests the direct method was modified by enrichment planting with spruce after removal of the standards and thinning of the coppice in each annual coupe of the old coppice cycle, thus effecting conversion in one period. In some of these forests Silver Fir seedlings were found, and this species was then planted instead of spruce, which being out of its habitat was not expected to seed well. As the conifers developed secondary and final fellings of the old coppice were carried out, and any seedlings of oak, beech, and ash were tended in order that mixed woods should be established.

Method of Conversion to Open High Forest (Futaie Claire)   In the north-eastern regions, particularly in Lorraine, when the blocks subjected to the light thinnings of the waiting-period method were due to be regenerated from 1903 onwards, there had been no good oak mast year since 1868. Only pure beech crops, or mixtures of hornbeam, aspen, birch, etc., could be expected. In 1905 Huffel, then Professor of Forest Management at Nancy, advocated thinning the whole area every fifteen years, because seed is produced in small quantities every year on oak trees which are given plenty of light. The increased light should also enable any seedlings which germinated to establish themselves, which he hoped they would do in groups. The thinnings in the upper storey removed all poor trees, any interfering with oak, and those causing local closing of canopy, while those in the underwood were heavy. A mast year occurred in 1912 and seedlings were established over very considerable areas which had been opened out. The method is now being applied in many forests of the north-eastern region.

In 1856 there were some 238,000 acres of coppice with standards under conversion: this rose to 740,000 acres by 1885, but fell to 230,000 acres in 1912. France has lost a great deal of revenue from the low timber production of coppice with standards, for which the lowered comparative value of wood fuel has not compensated, and many must have regretted that Lorentz's teaching was not heeded.

CHAPTER

22

In 1869 Karl Gayer became very alarmed by the state of
pure spruce plantations in Bavaria, as Cotta had been in
Saxony in 1844 (page 187). There was much damage by
frost in the young stands, by snow in the pole crops, by windfall in the even-canopied woods, and also by fungus and insect attacks. There was an accumulation of raw humus and
the crops developed an unhealthy appearance at an early age.
In other forests where the shelterwood strip system had increased the proportion of spruce in mixture with other species,
the resulting natural crops were relatively free from troubles.
Orientation of strips to avoid early morning sun on seedlings
had reduced frost damage, and with short cutting sections
producing only small even-aged areas, snow and wind damage
were not great, and insects did not cause serious injury.
Accordingly Gayer developed the group system, with fellings
proceeding against the wind. He made use of advance growth,
freed it, and obtained more regeneration in the groups before
weed competition became too great. When he wanted more
spruce and Scots pine he made his groups larger and extended them more rapidly than when he was favouring silver
fir and beech. Conversely, in crops in which light demanders predominated he made many small gaps, and opened
them up more slowly in order to give time for the desired
admixture of shade-bearing species to become established.

In the German-speaking cantons of Switzerland foresters
adopted Gayer's ideas, and stopped clear-felling in the older
mixed crops of the plains in favour of shelterwood strips with
short cutting sections. In 1874 the Confederation of Cantons
was formed and prohibited all further clear-fellings in the
mountains. In 1916 it was prohibited throughout Switzer-

land, including the private forests on the plains, because of the danger of the spread of insects from pure private plantations.

At the end of the 19th century Wagner did much work on the orientation of shelterwood strips to suit varying conditions of climate, soil, and aspect. He organized strip felling in great detail, prescribing modifications in the felling directions for various aspects, but compromised where it was necessary to avoid extracting timber through young regeneration. Some of his followers adopted his orientations without sufficient study of local conditions, so demonstrating once again the folly of copying the machinery of someone's idea instead of applying the idea to local circumstances and evolving the appropriate details from them.

On the eastern slopes in Germany storm damage generally started just below the crests and later the wedge system with the point of the wedge towards the wind just below the crest was tried. The advantage of the long cutting edges so provided was that cutting could be held up at points along them where regeneration was tardy, and they developed into irregular, wavy edges. Felling was done into the old woods through which the timber was extracted down the slopes. In Saxony the effects of repeated clear-felling and planting of pure spruce were becoming very evident. By 1920 the third rotation crops were well advanced in age, and measurements showed that annual production in the worst areas had dropped by as much as 50 to 60 cu. ft. per acre from that of the first rotation, less than 100 years before. Investigations suggested that the raw humus of pure spruce increased the acidity of the soil and exaggerated the effects of drought. Growth checks occurred after drought years in the pure plantations, but were much less serious in the remaining mixed forests in the mountains, and also in spruce plantations in which there was some natural beech. The need to lengthen the short rotation which had been fixed for the production of pit and pulp wood, to reduce the cut and to introduce mixtures was realized. Group fellings were made to allow beech to be introduced by planting, and the crops were thinned heavily to stimulate crown development and early seed production. Any advance growth found was freed and gradually a shelterwood group *cum* strip method was introduced.

These practices, which frequently involved very scattered regeneration operations, often over quite small areas, necessitated some revision of the age-class check, as previously applied by Judeich in Saxony. A factor known as the Geometric Average Volume per acre was calculated by estimating the final yield at rotation age of each stand of one-half of rotation age and over, summing these yields and dividing their total volume by the total area of the stands. The possible area for an annual area coupe was then investigated in five different ways as follows

$$(1)\ \frac{\text{Area of Felling Series}}{\text{Rotation}} \qquad (2)\ \frac{\text{Area of Oldest Age Class}}{\text{Regeneration Period}}$$

$$(3)\ \frac{\text{Area of two Oldest Age-classes}}{\text{Twice Regeneration Period}}$$

$$(4)\ \frac{\text{Area of Half Rotation Age and Over}}{\text{Half Rotation}}$$

$$(5)\ \frac{\text{Area of Stands listed for Regeneration on silvicultural grounds}}{\text{Regeneration Period}}$$

Each of these five possible coupes by area was then multiplied by the geometric average volume per acre to give five possible volume yields. A volume yield which was judged to be a suitable compromise was then decided, and the silvicultural annual coupe (No. 5 above) was adjusted so that it would give this yield.

The acceptance of scattered advance regeneration automatically increases the number of cutting points in regeneration operations, and so increases the difficulties of management. However, many foresters who had to deal with forests growing on relatively long rotations, in regions where regeneration was not always easy, came to realize that the periodic gift of seedlings by nature in various parts of a forest should not be lightly refused, simply for the reason that their incidence did not coincide with plans to regenerate specified crops in a particular period. In conditions such as are found in the Normandy forests, for instance, it is easy to pull up advance growth which may have been suffering from suppression for a time, and to obtain new seedlings just when they are wanted. Under less favourable conditions foresters often found that when they came to regenerate a scheduled stand they got nothing but weed growth, though some years before there had been a good stock of young seedlings which had been allowed to die because they did not form part of the plan.

In 1890, 1901, 1913, 1923, 1935, and 1948 there were exceptionally good mast years. This cycle of ten to twelve years coincides with a weather cycle which has not been explained. It has been observed that all tree species, when growing well within their climatic range, usually produce a phenomenal amount of seed at least once within this cycle. At each of these dates an increasing number of foresters in Switzerland, Germany, and France, set out to save as much as possible of the resulting regeneration, often in areas where they had experienced difficulties in obtaining natural regeneration to set time-tables, generally on account of excessive weed growth in regeneration areas waiting for seed.

In all cases other than those of forests managed by a selection system, special measures have to be taken if it is desired to save regeneration in areas not previously listed for regeneration at that time. In immature woods down to pole crops, group fellings, uniform fellings, or heavy thinnings are needed. To compensate for the intensification of fellings in the immature woods the fellings to free established regeneration in the regeneration coupes have to be reduced. This raises the principle of a long regeneration period and four important issues have to be faced.

(1) The effect of the retention of overwood in regeneration areas on the growth of the regeneration.
(2) The extent to which additional increment put on by this overwood compensates for any loss of increment on the regeneration.
(3) The extent to which the regeneration will be damaged by the eventual felling of the trees retained above it.
(4) The effect of the fellings made in immature woods to save advance regeneration on the capital and increment of the forest.

All the pioneers recognized that it was necessary to measure and remeasure the various crops and to keep careful records of the yields from each. Some measured all trees in every compartment, some measured sample plots only. Stephanie, of Baden, adopted the very practical procedure of measuring each year's coupe after the markings had been made and before the fellings commenced, keeping the figures for the proposed fellings separate from those of the trees to remain. In this way he spread the enumeration of each felling series over the whole felling-cycle.

The figures which they collected showed, according to their claims,

(1) That when dealing with oak, ash, and Scots pine, as well as with spruce, silver fir, and beech, increment check caused by leaving seed trees over regeneration was not serious for the established regeneration. Side light, particularly from the south in the northern hemisphere, was almost as effective as top light. Further, the young regeneration cleaned itself better if slightly checked, even if it were not dense, and so produced higher value increment in compensation for loss of volume increment.

(2) If vigorous trees were retained for the overwood, increased increment was obtained which was also of higher value, because of the large and well-shaped trees on which it was laid, provided that coarse trees were not kept.

(3) That careful felling by highly-trained personnel, who when necessary trimmed the crowns before dropping the trees, did surprisingly little damage. Definite extraction routes marked out through the regeneration and careful handling of logs prevented any serious extraction damage.

(4) In immature pole crops heavy thinning, by removing coarse, large trees and small bad ones, increased increment and value, and gave enough light to ensure the survival and slow growth of advance regeneration, even of such light-demanders as Scots pine. Where there were no good trees to leave it paid to cut out poor ones, and so to allow the advance regeneration to develop and make the canopy uneven, which diminishes wind damage. While making these thinnings mixture by groups could be obtained for the next crop.

Foresters who have set out to save advance regeneration all over their forests have proceeded to subject the resultant crops to a system of management. This is essential, because while management which is not based on good silviculture definitely harms a forest, silviculture which is not organized in a manageable manner breaks down and leads to nothing but confusion. One particular forester with a skilled and enthusiastic staff may be able to work a small forest on purely silvicultural conceptions without any definite scheme. When he, with his knowledge of all the individual plots, leaves it, no one else can carry on. When a large forest is concerned practical planning is essential, or masses of small groups which need attention will be lost sight of, or will not be tended at the right time, because there is no workable scheme to ensure that it is done.

In Switzerland the systems adopted have been the Baden and Swiss *Femelschlag* (method of progressive regeneration fellings with a long regeneration period or, as it is sometimes called, the irregular shelterwood system), and the modification of the selection system known as Biolley's *Méthode du Controle*.

In the Swiss and Baden *Femelschlag* system there are no periodic blocks: compartments and sub-compartments are grouped in cutting sections, as many sections in each felling series as there are years in the felling-cycle, which is usually ten or less. All the operations required in each cutting section are carried out on the felling-cycle, except the first increment thinning in the young pole crops. This is deferred pending a good mast year, or the appearance of advance growth, particularly when an invasion of rank weeds is likely to follow any breaking of the canopy. The regeneration areas are selected every ten years. They are usually groups of advance growth, and should be sited so as to avoid the need for extraction through established saplings later on. The length of the regeneration period is about half the nominal rotation: nominal because a number of the best trees are selected to remain standing as long as they continue to put on valuable increment. In each cutting section there will be, in time, uneven-canopied young woods, from which all the overwood has been removed; pole crops subject to increment thinnings, with advance growth developing in groups below them; and immature timber crops, which in various stages may look like irregular two-storied forest or even selection forest, but which, however, even up in a remarkable manner in the course of the removal of the overwood. The sequence of the regeneration is, generally, first beech, then under it, but soon pushing through and tending to dominate it, silver fir, and, as heavier fellings are made, spruce between the groups, joining them up. Some Scots pine are often left standing to the last to seed up gaps caused by the final felling, or planting may be resorted to for this purpose.

One hundred per cent enumerations are carried out every ten years, and until recently the yield was calculated according to Heyer. Now the increment is usually calculated according to Biolley, and Heyer's formula as modified by Huber is used as a check. This is

$$\text{Annual Yield} = \text{Actual C.A.I.} + \frac{\text{Actual G.S.} - \text{Normal G.S.}}{\text{Period chosen for adjustment,}}$$

in which the second expression may be a minus quantity. The actual annual increment is ascertained by measuring the periodic annual increment of each stand for the last ten years, adding these together and dividing by ten. The normal growing stock is calculated according to Flury as $I \times cR$ (in which c is the appropriate empirical constant for the composition and rotation of the crop in the conditions existing locally). The I used in the calculation is the sum of the current annual increments of each age-class, adjusted to its normal and normally stocked area.

An area check is also made to ensure that in each ten-year period regeneration is being carried out over approximately the right-sized area; for instance, a tenth of the whole if the rotation is a hundred years. The method is an intensive and elastic one, well suited to the regeneration of mixed woods consisting mainly of shade-bearers. It needs very skilful handling and an intensive network of roads to permit of extraction from the many scattered felling points. It would be quite useless to attempt it where felling cannot be done by skilled woodsmen employed by the officer in charge. It is sometimes used as a transition system when woods are to be converted from an even-aged to an all-aged condition.

CHAPTER

23

A great deal has been written about Biolley's method of control in recent years. As a complete application of the selection system it can only be applied to forests of species which thrive in an all-aged composition, but the method of yield control, or rather of yield prediction, which he devised as part of it can be used with any form of forest treatment, except those involving clear felling on an area basis.

Henri Biolley, who studied forestry in Zurich, 1876–9, had been strongly influenced by the writings of Gurnaud, who had left the French forest service because he was not allowed to practise the selection system in the Jura forests. In 1880 Biolley took over the circle of the Vals-de-Travers, in his own canton of Neuchatel, in the Swiss Jura. He studied the forests, which were then in process of conversion from rather indiscriminate selection fellings to the uniform system with strip regeneration. By 1890 he had worked out in detail a system which he began to apply to the Forest of Couvet, which was 344 acres of silver fir, spruce, and beech. Later he extended the system to all the 7,500 acres of the canton, of which he became inspector in 1917. He retired in 1927 and his successors, Favre senior, 1928–46, and Favre junior have carried on his method exactly as he intended ever since. The success he achieved led to the extension of his system to all State and communal forests of the canton of Neuchatel and later to many other cantons. Biolley was obsessed by the idea that trees of all sizes growing together all over a forest can make better use of all levels of soil for water and salts, and of all levels of atmosphere for carbon than can even-aged stands on different parts of it. He likened an even-aged forest to a terrace of houses in which only one storey of each house is occupied at a time. He also wished to measure the increment of his forest, rather than to calculate it from formulæ. Biolley rejected all previous ideas of rotation, exploitable size, and normal growing stock distribution, and decided to leave all well-shaped, vigorous trees to grow on as long as they remained vigorous and did not harm potentially better replacements. He regarded the forest as a laboratory and management as the study of results obtained in that laboratory. 'Management,' he said, 'is the critic, councillor, and servant of silviculture,' and he thought that a working plan should sanction the putting into practice of periodic conclusions drawn from experiments by means of frequent revisions.

By 1890 Biolley had divided Couvet, Series I (136 acres), into thirteen compartments. These he enumerated individually down to a minimum diameter of 17·5 cm. (say, 7 inches) at breast height in 5 cm. (say, 2 inches) classes, marking the point of measurement on each tree with a scribe. He grouped his classes as follows (when quoted approximately in inches):

Group	2-inch diameter classes	Size limits of Group
Small	8, 10, 12	7 to 12·9 inches.
Medium	14, 16, 18, 20.	13 to 20·9 inches.
Large	22 and over.	from 21 inches upwards (possibly some of 45 inches).

Starting on crops which were all-aged in some places but contained some even-aged strips and groups he cut out poor trees and began the long task of adjusting the crop mixture in each compartment to his likiing. He worked on a six-year felling-cycle, sometimes leaving a compartment for seven or eight years, covering two or three compartments a year. After 1932 the felling cycle became seven to eight years. One hundred per cent enumerations [1] were repeated every six years (seven years after 1932), and the volume of the growing stock at the second enumeration plus the volume removed in the intervening period, minus the volume at the first enumeration, gave the volume of growth during the period. That is to say, if I represents the total increment during the period, N the volume of the cut, and V_1 and V_2 the measured growing stock at the beginning and end of the period respectively, $I = V_2 + N - V_1$. Biolley recognized two possible sources of error in the calculation.

(1) The volume cut must be measured in exactly the same

[1] Taking about twelve man-days of work per 100 acres; cost about 3s. 4d. an acre at 1951 rates.

way as are the standing trees, i.e. by a volume table and not with tapes on the ground. Special volume tables are used, sometimes simple volume tables based on the average height curve for a particular forest, or compound tables based on the height curves for different species and productivity sites in the forest, or sometimes a general table for a fairly large area such as a canton. Agreement between the volume read from the table and the actual measured volume on the ground is not of vital importance if the two volumes are never confused or used in combination. To ensure against this Biolley adopted a special term for his volume table unit, and called it a 'silve'. The silve is, of course, approximately a cubic metre, to which its relationship can be determined at any time by measuring on the ground a number of felled trees already measured standing by the table. The relationship varies slightly from compartment to compartment, and also in the same area from time to time as the taper of the stems increases or decreases. The ratio cubic metre/silve has varied in Couvet between 0·97 and 1·00. Not all cantons make this distinction, and in some of the larger forests the output is measured on the ground. There are always possibilities of variation in measurement of tops for firewood, overbark, and underbark volumes, etc., if this is done.

(2) The increment put on by the previously enumerated trees must be kept distinct from the volume of new recruits to the lowest measurable class recorded at the second enumeration. This should not be important in a truly all-aged forest, in which recruitment should be roughly the same in all periods; but when converting from a partially even-aged forest, as Biolley was, whole strips or groups may attain enumeration size in one period, and none in the next. For this reason, and to ascertain the growth rates of the different size-groups, the stand table is checked off in descending order from the largest to the smallest diameter class, and the number of stems in each group is made to agree with the original number of the group as in the following table. The figures are hypothetical and chosen for simplicity. Only three diameter classes are shown in the Large Group, instead of the twelve to fourteen classes which would probably be present, in order to avoid overloading the table.

INCREMENT CALCULATION BY DIAMETER CLASS GROUPS

Group	Diam. Class, in.	G.S. 1900 V.1 No.	G.S. 1900 V.1 cu. ft.	G.S. 1906 V.2 No.	G.S. 1906 V.2 cu. ft.	Fellings 1901–1906 N. No.	Fellings 1901–1906 N. cu. ft.	V.2 + N. No.	V.2 + N. cu. ft.	Reconstruction of V.1 No.	Reconstruction of V.1 cu. ft.	Recruitment No.	Recruitment cu. ft.	I. in 6 years	I. in 1 year	1% per annum on V.I
Large	26	8	1,200	9	1,350	3	450	12	1,800	12	1,800					
	24	9	1,080	10	1,200	1	120	11	1,320	11	1,320					
	22	11	1,155	12	1,260	—	—	12	1,260	5	525					
		28	3,435	31	3,810	4	570	35	4,380	28	3,645	7	735	$\left\{ \begin{matrix} 3,645 \\ -3,435 \end{matrix} \right.$	$\dfrac{(210)}{(6)}$	$\dfrac{(35 \times 100)}{(3,435)}$
														210	35	1·02%
Medium	20	12	1,020	12	1,020	1	85	13	1,105	7	735					
	18	14	910	14	910	1	65	15	975	13	1,105					
	16	18	900	15	750	4	200	19	950	15	975					
	14	16	560	16	560	2	70	18	630	19	950					
										6	210					
		60	3,390	57	3,240	8	420	65	3,660	60	3,975	12	420	$\left\{ \begin{matrix} 3,975 \\ -3,390 \end{matrix} \right.$	$\dfrac{(585)}{(6)}$	$\dfrac{(97·5 \times 100)}{(3,390)}$
														585	97·5	2·88%
Small	12	22	550	20	500	3	75	23	575	12	420					
	10	38	570	32	480	5	75	37	555	23	575					
	8	50	500	45	450	11	110	56	560	37	555					
										38	380					
		110	1,620	97	1,430	19	260	116	1,690	110	1,930	18	180	$\left\{ \begin{matrix} 1,930 \\ -1,620 \end{matrix} \right.$	$\dfrac{(310)}{(6)}$	$\dfrac{(51·66 \times 100)}{(1,620)}$
														310	51·66	3·13%
Total		198	8,445	185	8,480	31	1,250	216	9,730	198	9,550			1,105	184	2·18% Increment
										18	180			180	30	0·35% Recruitment
										216	9,730			1,285	214	2·53% Total Inc.

New Stems

Check of $V_2 + N$ 216   9,730

In Couvet this calculation was made for each compartment separately every six years from 1890 to 1932 and then every seven years, and the volume to be cut from each compartment for the next period was decided after consideration of the condition and distribution of the growing stock and the increment rates of the various size-groups in the past. In some other forest districts the calculation is made for the district as a whole and the yield prescribed for the next period is divided up among the compartments in the plan of exploitation. In some cantons which have forest areas up to 17,000 acres ten-year enumeration and working-cycles have been adopted, the silve is not used and no attempt is made to distinguish between increment of enumerated stock and recruitment.

However carefully measurement and remeasurement are carried out the possibility of errors in enumeration and calculation for small areas at short intervals are such that no very far-reaching conclusions can be drawn from the figures for any one period, or from comparison of the figures for any two periods. After several enumerations, each acting as a check on the others, this difficulty disappears and any misjudgement of what the yield should be is soon reflected in changes in the growing stock. It is recognized that compartments used as units of calculation should not be smaller than $12\frac{1}{2}$ acres, and that a size of 25 acres is probably safer.

The success of the method depends on the skill with which the marking officer, who must know his forest intimately, harvests the yield decided upon. On entering a compartment he first looks to the ground to ascertain if there is ample seedling regeneration of the desired species. Then he looks to see if there are sufficient saplings becoming established for selection in the choice of an adequate number of good poles. If he is satisfied on these points the overwood cannot possibly be excessive: if not he must take the yield from the trees casting most shade. He then looks higher to decide if, after he has cut poorly-shaped and wolf trees, there will be enough first-class poles to make the timber trees of the future. Then he must decide if there are enough good immature trees over two-thirds of the lowest diameter for valuable timber to permit the removal of some of the largest trees. Among such trees he looks for ones which are interfering with other large trees and at the same time holding back promising smaller ones. Trees to be left are selected for their vigour, as shown by the colour of the crown and cleanness of bark; increment borers are never used. The aim is to leave large well-shaped trees to put on as much increment as they will.

As in the *Femelschlag* system felling is done in the winter with great care by skilled fellers. In Couvet Series I the forest guard fells each tree and with it drives into the ground a peg placed on the line where it will do least damage. Sometimes the larger branches are cut off before felling, and always cultural operations are carried out in the cut-over areas in the following spring.

Extraction is done over snow on carefully selected routes to the excellent network of roads.

It has been found that climatic variations in the different periods have a marked effect on growth, and it is hard to differentiate between these and the effects of the size and size-class distribution of the growing stock. As previously stated, it is the general trend of changes that has to be watched and compared with what is happening in other forests. The figures below are for Series I at Couvet and the increments quoted include recruitment to enumeration size as being the fairer index of growth over a long period. For conversion to cubic feet it has been assumed that a silve equals a cubic metre.

Period	Growing Stock, cu. ft. per acre	Increment, cu. ft. per acre per annum	Percentage by Volume of Growing Stock in the Size-groups		
			Small	Medium	Large
1891–96 .	5,600	119	24	49	27
1897–1902	5,430	116	22	48	30
1903–08 .	5,300	139	20	47	33
1909–14 .	5,260	162	17	45	38
1915–20 .	5,200	129	14	42	44
1921–26 .	5,050	119	12	40	48
1927–32 .	4,900	124	12	38	50
1933–39 .	4,820	172	12	35	53
1940–46 .	5,180	104	12	31	57
1947 .	4,990		14	28	58

Falling off in increment similar to that shown above in the periods 1915 to 1926 has been found by stem analyses of trees in other forests of the French and Swiss Jura and in the Swiss Midlands. Probably it will be found to have occurred all over the area in the period 1940–46 also.

What is really striking about the figures is the steady increase in large timber which has meant higher financial returns. The proportion of saw timber in the yields has risen from fifty-six per cent in 1890 to eighty per cent. The net money yield per acre per annum rose by 600 per cent between 1881 and 1939, some of which is, of course, due to the rise in the price of timber, but the cost of labour has risen also and the main cause is the higher yield of the larger sizes and better grades of timber. (Annual recurrent expenditure is about twenty-five per cent of the gross annual income.)

What appears to have happened is equivalent to an increase in the rotation in an even-aged forest without the inevitable decrease in area of the oldest age-class in such a forest. As individual trees are kept standing just as long as they continue to grow well, all that is left of the idea of a rotation is an average age of exploitability which varies from period to period, and there can, therefore, be no constant equivalent normal growing stock or age-class distribution. According to the theory of shared space an all-aged forest can carry a larger growing stock than an equivalent area of even-aged stands. The figures from Couvet do not demonstrate this, possibly because the process of bringing a partially even-aged forest into an all-aged state required a preliminary reduction of stock. They do show, however, that the proportion of the older age-classes can be higher. The intention now is to seek an *étale*, which can be rendered as a provisional limit of capitalization. In other words the largest possible stock of large timber compatible with an all-aged condition will be built up, and if, allowing for climatic variations, a point is reached at which the larger stock gives a smaller increment, the stock will be reduced again. No theories, but experiment and application of the results of the experiments will guide the management.

In the Schwarzenegg district forests which have always been managed by selection carry stocks ranging from 7,000 to 8,500 cubic feet per acre and increments from 72 to 185 cubic feet per acre per annum have been recorded. Stocks, however, are being reduced in order to maintain the all-aged character of the forests.

The publicity which Swiss selection forestry has received must not blind us to the excellent results which have been obtained in spruce and silver forests growing in similar conditions in the French Jura from management as uniform forests. La Fuvelle, a forest of about the same size as Couvet at much the same altitude but with a rather higher rainfall, still has the permanent periodic blocks laid out in 1858. In 1938 the growing stock was 6,200 cubic feet per acre, and the average production per acre is about 129 cubic feet a year.

The large forest of La Joux, managed by Mélard's regeneration area method, is at a rather lower altitude and enjoys a moister climate than Couvet. For the period 1910–26 it had an annual increment per acre of 175 cubic feet from an average growing stock of 6,120 cubic feet per acre. Both these forests are, of course, skilfully handled by foresters who know them well, and one is forced to the conclusion that, in circumstances favourable to tree growth any carefully organized system, which takes into account the needs of the species and all local factors, can be applied successfully by a good forester.

In some French forests in which yields are regulated by area and volume, it is a common practice to calculate the increment put on during a period. To do this the growing stock at the end of the period and the cut during it are added together, and from their sum the growing stock at the start of the period is subtracted. The result is divided by the number of years in the period and expressed as a percentage of the mean of the two growing stocks (see page 127). This percentage applied to the present growing stock is used as a check on the yield calculated by Mélard's method. The same volume table (usually a quasi-conventional one) is used for the calculation of the volume at each enumeration, with the avowed intention of obtaining comparable results. Some foresters have pointed out that the ratio of the volume-table volume to the real volume may well have altered in the interval. If the stands of the series as a whole have aged, the ratio will have risen, and if they are younger on the average, the ratio will have fallen. In all-aged forest for which Biolley devised his method this difficulty should not arise.

CHAPTER
24

A reminder of the impact of human activity on forests and of the familiar theme that forest practices evolved to suit one set of conditions may be dangerous in other conditions can be provided by a brief glance at Norway.

Forests cover some 20,000,000 acres or twenty-five per cent of Norway, and much of this land could not produce anything else of value. Export of logs was started by the Vikings, and by the 13th century had become a paying industry involving river floating. Enterprising private persons sought and obtained grants of eighty-five per cent of the forests and water-powered sawmills were started in the 15th century. Several attempts were made to restrain forest destruction and a forest service was started in 1739. This existed for only seven years, was revived in 1760, but still public opinion was not ready and it was killed again in 1771. Shortly after this commercial clamour succeeded in having all restrictions on private trade in timber removed. By the middle of the 19th century the effects of this were causing real alarm and the present forest service was founded. German ideas that natural forests consisted of trees of all ages and were best managed by a selection system were adopted, and in 1863 control of cutting by the imposition of minimum cutting diameters was introduced. However, in rigorous northern climates the soil temperatures in coniferous forests are too low for any breaking down of their acid litter. Left to themselves they gradually deteriorate till fire or storms sweep them away, expose the ground to the sun, and permit the formation of humus. In 1837 a heavy storm cleared many square miles of old depreciating forest and natural regeneration occupied the ground. After this clear-felling methods by strips, etc., were tried and proved to be successful. In 1932 all private forest owners were required by law to adopt good forest management, which was defined as meaning that fellings in young stands must not be made for the purpose of producing yield, and in old stands must not prevent regeneration. In all managed forests management was changed from the selection system to stand management.

In 1930 a forest inventory taken by means of sample strips all over the country disclosed that the average growing stock per acre was 600 cubic feet, and the average increment per acre per annum was 20 cubic feet. In 1950 a second inventory showed that the average growing stock had increased by twenty per cent and that increment per acre per annum had increased to 28·5 cubic feet.

Relationship between Normal Growing Stock and Production during a Rotation of seventy years on a Normal Series of Plantation Teak, Site Quality I (based on the Yield Table of Laurie and Ram, Indian Forest (New Series), Silviculture, Vol. IVA, No. 1, 1940).

Age, years.	Main Crop, cu. ft.	Thinnings, cu. ft.	Total Production, cu. ft.	M.A.I., cu. ft.	C.A.I., cu. ft.	Final Yield, cu. ft.
					o	
5	o	o	o	o		o
					68	
10	340	o	340	34		340
					74	
15	690	20	710	47		710
					94	
20	1,080	80	1,180	59		1,160
					114	
25	1,490	160	1,750	70		1,650
					138	
30	1,910	270	2,440	81		2,180
					160	
35	2,350	360	3,240	93		2,710
					176	
40	2,810	420	4,120	103		3,230
					180	
45	3,280	430	5,020	112		3,710
					174	
50	3,740	410	5,890	118		4,150
					168	
55	4,200	380	6,730	122		4,580
					154	
60	4,650	320	7,500	125		4,970
					144	
65	5,110	260	8,220	126		5,370
					126	
70	5,740	—	8,850	126		5,740
		3,110			1,770[1]	40,500
						− 2,870[2]

On 14 acres in 1 year Production = 1,770,   N.G.S. = 37,630
On 70 acres in 1 year Production = 8,850,   N.G.S. = 188,150

Note 1   The C.A.I. figures can be regarded as those for 14 sample acres, each representing 5 acres, and can be added direct because they are the true averages for each period, e.g. 126 cu. ft. are put on each year on each acre from 66 to 70 years old.

Note 2 The final yield of 5,740 cu. ft. cannot be taken as the average on 5 acres 66 to 70 years old, but only of $2\frac{1}{2}$ acres $67\frac{1}{2}$ to 70 years old. The average at 5 years is the average on 5 acres $2\frac{1}{2}$ to $7\frac{1}{2}$ years old, and the $2\frac{1}{2}$ acres 1 to $2\frac{1}{2}$ years old also carry no timber.

Annual Production from Series = Final Yield 5,740 cu. ft., 65 per cent
 + Thinnings 3,110 cu. ft., 35 per cent

Total Production from Series of
   70 acres each year = 8,850 cu. ft.
Total Production from Series of
   70 acres in Rotation of 70
   years = 619,500 cu. ft.

i.e. starting with a Normal Growing Stock of 188,150 cu. ft., in 70 years 619,500 cu. ft. can be cut and 188,150 cu. ft. will remain as a Normal Growing Stock.

Hence, Yield during Rotation $= \dfrac{619,500}{188,150} = 3 \cdot 3$ times Normal G.S.

Comparison with the theoretical conception that,

$$\text{N.G.S.} = I \times \frac{R}{2}$$

$$= 5,740 \times \frac{70}{2} = 200,900 \text{ cu. ft.,}$$

and that Yield during Rotation $= 2$ G.S. (or G.S. $+ \frac{1}{2}(I \times R)$)
         $= 401,800 \; (200,900 + 200,900).$

In this case the Flury constant can be calculated as,

$$c = \frac{\text{N.G.S.}}{I \times R} = \frac{188,150}{5,740 \times 70} = 0 \cdot 46827$$

and will naturally give the correct N.G.S. on 70 acres from N.G.S. $= I \times CR.$

APPENDIX II

ORGANIZATION OF SUSTAINED YIELD IN PREVIOUSLY UNMANAGED FOREST

(*Paper presented to the British Commonwealth Forestry Conference, Canada, 1952.*)

For periods within the physical rotations of the species which comprise them, mature, unmanaged forests can be assumed to be in a state of dynamic equilibrium. Large, old trees are generally in excess of medium-sized and small trees of the same species, which only have an opportunity of entering the top canopy as overmature trees die. Because no increment is harvested it has slowed down until it is practically balanced by decay. The smaller trees are not all necessarily younger than the large ones, but may merely be suppressed and even incapable of responding to release. Abundant seedlings are liable to appear on the forest floor after mast years, but most of them fade away in a few years' time, leaving only individuals to survive in gaps here and there.

When such a forest is opened for exploitation the primary aims of wise management are to harvest each year from the existing growing stock a regular volume which will be economic to cut, and meanwhile to procure regeneration and bring it to maturity, so as to enable the same, or pre-ferably, a larger volume to be cut annually for ever. To achieve this forests are grouped or divided into production units, each estimated to be capable of providing permanently the raw material needed for a plant, such as a sawmill, or a combination of plants for different classes of material, established to process it.

There are, of course, cases in which the produce from small forests can be sold at any time it is available to markets which are not dependent on it, but are fed from numerous sources tapped by highly-developed trans-port routes. Such woods do not necessarily provide regular incomes to their owners or security to a wood-using industry, for both of which purposes sustained yield management is essential.

A previously unmanaged forest may or may not have been exploited in the past. Some will have been subjected to selective felling or 'creaming' of the best trees, which, if continued, will reduce or destroy for a long time their economic value. It is essential that management should be started while the volume and value of merchantable timber which can be cut per acre are sufficient to make extraction profitable in the parti-cular locality. Also there must be available a sufficient acreage to provide a regular supply of material for a plant of economic size, during the period which will be required to replace the existing trees by a balanced crop capable of maintaining the supply. The manner of replacement, the most suitable arrangement of age-classes in the new crop, the time which it will take to reach exploitability, and the prospects of supplying part of the yield during the replacement period from trees not yet of economic size, will all require investigation.

225

It is sometimes tempting to think, particularly in tropical forests containing a sprinkling of big trees of valuable species and some smaller ones, that selective felling might be controlled and developed into true selection working. In order to harvest the largest trees before they deteriorate, the whole forest would have to be worked over on a short felling-cycle with a high minimum girth or diameter limit for felling, and then again with a lower limit, and so on till forest all-aged up to the desired exploitable size resulted. Unless the species were of great value such scattered exploitation would be completely uneconomic. The selective felling referred to was not the removal of the most overmature trees, which were often mis-shapen and so were left sterilizing space, but of the best trees. Even so, it only paid in the richest and most accessible parts of the forests in which it was practised. The gaps which were made in the canopy often filled up with weed species, and frequent tending would obviously be needed to establish saplings of the valuable species in them; again a dispersed operation which would prove expensive both in supervision and labour.

True selection forestry with continuous regeneration can only be carried out with fairly shade-tolerant species, and requires intensive control by highly skilled and experienced foresters with full silvicultural understanding of the species with which they are dealing. It also needs meticulous care in the felling and extraction of timber, which must be carried out by specially trained operators who are more interested in preventing damage to the trees which remain than in their own rate of output. Group selection is only one degree less difficult and diffused, and needs as much skilled supervision. Selection properly applied involves returning at intervals of not less than ten years to the same coupe to remove only those trees whose growth has culminated since the last visit, to release promising younger trees from any threat of suppression, and to tend saplings; so that no increment is lost by degrade, growth never slows down and regeneration goes on all the time.

According to the theory of shared growing space, such a forest can carry a greater volume of large trees relative to smaller ones than can a uniform forest and, as proved in comparatively small and highly staffed woods in Europe, can produce a large sustained yield of high-quality timber. Where low stumpage values necessitate concentrated working, mass extraction by heavy machinery, a minimum of permanently maintained roads and economy in trained staff all-aged forest cannot be built up. Management by selection may be likened to production in a craftsman's workshop and management by a uniform system to production on an assembly line, where semi-skilled operators can be employed. It is therefore probable that replacement by even-aged stands will be chosen as the immediate objective when an unmanaged forest is first organized.

In order to convert a forest in dynamic equilibrium into a series of uniform stands which will provide a sustained yield, determined by the rate of growth obtainable in the edaphic and climatic conditions present, the old crop has to be cleared, either coupe by coupe in one or two operations, or progressively over a series of coupes organized as a periodic block, as new seedlings are established. Unless the forest is very well stocked and readily accessible, it is unlikely that felling the existing merchantable trees on a coupe or periodic block in several operations will be economic. Sometimes there will be present on the ground adequate numbers of seedlings of desirable species which, though normally they would disappear when a few years old, can be established by opening the canopy. Sometimes it is possible to induce such seedlings by cheap pre-exploitation treatment, such as in the tropics, climber cutting and poisoning of useless trees. Sometimes a certain number of seedlings will appear after heavy exploitation from seed already in the ground, and from that provided by trees of the valuable species left because they are below the economic cutting size, but these may have to be supplemented by artificial planting. In other cases it may be necessary to clear-fell in strips or plots which can be seeded from adjoining forest, or in blocks to be planted artificially. In tropical forests the usable species generally form only part of the growing stock and, after cutting all that can be sold, partial shade is left, which may be made use of to nurse regeneration of valuable species, or which may have to be cleared and burnt prior to planting, possibly by forest cultivators. The most promising method of restocking the coupes to be exploited will have to be chosen, provided, of course, it does not place restrictions on exploitation which would render the latter uneconomic. If it does so, no cutting would take place at all and the forest would remain unproductive. This position may sometimes have to be accepted temporarily when the revenue that would be obtained from working would not pay for the only method of regeneration likely to succeed. In such a case the forest should be left alone until local stumpage values rise, additional species become marketable, or research provides a cheaper method of regeneration.

When the condition of dynamic equilibrium exists, it would appear that the sustained yield during the conversion period would be the present volume of trees of merchantable species which have attained merchantable size, and that the conversion period must be long enough to allow the new crop established on the first coupe exploited to reach exploitable size. This overlooks the possibility of harvesting trees which, though below economic size when a coupe is exploited, may respond to release and reach such a size in less than the conversion period. In many cases such harvesting will not be economic, particularly when account is taken of the damage which will be done to the smaller trees during the original exploitation. Conversion in one felling-cycle, equal to the future rotation of the regeneration to be established, must often be accepted as the only safe way of assuring sustained yield.

The simplest method of conversion is to divide the whole area by the number of years in this felling-cycle or conversion period, and to fell and regenerate one of the equal annual coupes so determined each year. Unless the forest is exceptionally homogeneous this will not provide equal annual yields. All forests are likely to cover sites of varying productivity and to contain portions which have been damaged by storms, fire, or other agencies. In the tropics there are often several crop types

227

present, each with a different proportion of merchantable species. If a sufficient acreage is available to allow a portion to be kept in reserve to supplement the poorer coupes, steady provision of produce can be maintained by drawing on this reserve, but then area control has broken down, and there is no basis of regulation except that of the volume which may be cut each year. The alternative of laying out equi-productive coupes before exploitation may be ruled out as too laborious for large tracts of little-known forest. Conversion with volume control requires assessment of the merchantable volume standing on a production unit before any felling is started, but without such assessment the capacity of a processing plant to be fed from the unit cannot be determined. An elaborate enumeration survey is not essential, because the information required is not so much a precise figure of the volume which will be available each year as a safe forecast of the annual minimum. If this minimum is exceeded in practice no great harm will be done, but if it is not reached the plant cannot be operated economically. Therefore an estimate based on a low intensity of sampling, with a generous allowance for defective and badly shaped trees, and the deduction of a percentage as a safety reserve, should suffice for the preliminary calculations. The estimate can be improved at each revision of the working plan according to the actual results obtained in working, and provided that the original figure was a safe minimum, it will be easy to increase the prescribed annual yield and to expand the processing plant.

The annual yield fixed will be simply the estimated volume of timber of merchantable size and species in the forest, less the safety reserve, divided by the conversion period, because no increment can be expected to accrue on any part of the area before it is exploited. This yield is harvested from annual coupes by laying out three sides of a coupe and exploiting it systematically in contiguous strips from side to side, so that felling proceeds as far towards the open end as is necessary to obtain the annual volume, and the coupe is then closed. This really amounts to laying out equi-productive areas during exploitation. In some cases the yield so controlled will be that of a group of valuable species only, such as Dipterocarps or *Meliaceae*, and other species may be cut within the coupe during the year at the discretion of the officer in charge, who may wish to retain a certain density of canopy.

There are, however, many types of unmanaged forest in which, possibly because they are not fully mature, or because of the shade tolerance of the more important species, age class distribution is definitely good. Such forests might well be managed under a selection system if diffused felling would be economic in them, and if adequate staff with skill and experience were available for both the silviculture and extraction. In them it is not only definitely wasteful to leave the trees of merchantable species not yet of economic cutting size to deteriorate over the full conversion period, which is a rotation for the species, but also detrimental to the growth of the new uniform crops to be established beneath them. If evidence can be obtained that such trees will respond to release, consideration must be given to the possibility of harvesting those which escape destruction during the first exploitation, at a later stage in the conversion. The question is whether after at least an economic volume per acre has been cut in a felling-cycle shorter than the conversion period, the residual trees will provide an economic volume per acre again during a second cycle equal to the balance of the conversion period. To determine this information must be obtained as to the percentage of the residual trees of various sizes which will escape serious exploitation damage and respond to release, and their probable rates of growth. Once again, great accuracy is not essential as long as the figures adopted are conservative. For instance, when nothing better is available, such generalizations as that many tropical species will average 1 inch of girth increment in one year, or 3 inches of diameter increment in ten years, can be adopted. The extent of felling damage and the survival prospects of the various size-classes can be based on observations in similar forests which have been exploited. Enumerations will have to be extended below the minimum economic limit for utilization; in fact, the smallest size about which information will be required is that which will attain this minimum limit in half the conversion period.

This is because, whatever the lengths of the two felling-cycles into which this period is divided, the average number of years which will elapse between cuttings in the same area, will be half the sum of the two cycles. The first coupe will be exploited again at the beginning of the second cycle and the last in the last year of that cycle.

What cannot be calculated, but can be taken into account as a safety factor, is that, after the whole forest has been worked over once, the extraction routes will be less expensive to open for the second cycle. Also by the time the second cycle is started stumpage values will probably have increased, making it economic both to cut smaller trees and a smaller volume per acre; also possibly a greater number of species. This might justify the adoption of two cycles in marginal cases. There will inevitably be some exploitation damage to the established uniform regeneration during the second cycle, but this should be offset, to some extent at least, by release from suppression and the opportunity to carry out a tending operation. In large tracts where stumpage values were low, it would not be possible to afford tending operations in regenerated coupes once the young trees were established, unless revenue would accrue from such operations. Extraction of thinnings without some mature trees would not be a paying proposition, and in many forests the organization of a second cycle on the basis of an economic cut per acre is the only way in which entry into regenerated coupes during the rotation of their crops can be made possible.

In the unlikely event of the residual trees left during the first felling of an annual coupe being able to attain in half the conversion period the same volume as was cut originally from the coupe, each felling-cycle could be half the conversion period, and the average area required to provide the same annual yield in each cycle would be the same. If the residual trees would attain half the volume cut originally the second cycle would have to be half the length of the first, and the average annual coupe in the second cycle twice the area of that in the first. Thus, the number of years in the two felling-cycles must bear the same proportion to each other as the volume of exploitable timber per acre in the first cycle bears to the estimated exploitable volume per acre in the second cycle.[1] If the lengths of the two cycles are properly adjusted, the annual yield throughout the whole conversion period will be the estimated total volume of exploitable timber on the whole area at the beginning of the period, less a safety percentage reserve, divided by the number of years in the first cycle only. The whole yield in the second cycle will come from trees below the economic felling limit at the start, which are entirely ignored when conversion is carried out in one cycle only.

In certain circumstances it may be possible and desirable to compromise slightly on the question of the minimum felling size for the first cycle. If the forest is well stocked with large trees, the limit might be raised so that it still permits of exploiting an economic volume per acre, but leaves rather more residual trees to provide the yield in the second cycle. This will have the effect of shortening the first cycle and so allowing the cutting of a larger average annual coupe which will compensate the exploiter for his sacrifice in accepting a high exploitation limit. The total yield obtained from the forest in the conversion period should be greater than that obtainable with the lower limit, provided that the increment on the additional trees left is more than the loss by felling damage and mortality among the greater number of residual trees.

CONCLUSIONS

It appears that the more gradually the large trees in a previously unmanaged forest can be cut, the greater will be the sustained yield during conversion to fully stocked forest containing all age classes in the proportions needed to provide the maximum sustained yield permanently. How gradually they can be cut depends on the richness of the forest, stumpage values in the particular locality, the ease with which internal communications can be constructed and maintained, the quality and quantity of the staff available for management, and the methods of logging which can be adopted. The annual extraction in an economic manner of an adequate volume, which will cover the cost of establishing regeneration in the area worked over, is essential to enable conversion to be undertaken at all. At the one extreme of a remote and roadless tract this may necessitate concentrated felling of all trees of merchantable size and the writing off of all smaller trees and their potential increment. At the other extreme of a forest with a network of internal roads and rides

[1] If a = vol. per acre of exploitable timber now, b = vol. per acre expected after half the rotation, x = years in 1st cycle, y = years in 2nd cycle, $x + y$ = rotation, and $x : y : : a : b$, then

$$\frac{x+y}{x} = \frac{a+b}{a}, \; x = \frac{a(x+y)}{a+b} = \frac{x+y}{\dfrac{a+b}{a}} = \frac{\text{rotation}}{1 + \dfrac{b}{a}}$$

situated in the midst of a well-developed area, it should be possible to cut all the existing trees progressively as they reach maturity, and so to harvest the maximum of increment. There are many forests between these extremes which it should be possible to convert in two felling-cycles, harvesting some of the potential increment of the smaller trees, while still maintaining economic conditions for commercial working.

As this two-cycle method is not one which is in general use a summary of the steps required to apply it and examples from two forest types of the calculations required are given below.

METHOD OF APPLICATION

(1) Select a rotation suitable for the uniform crop to be established, and adopt it as the conversion rotation.

(2) Enumerate the growing stock by convenient size-classes down to a minimum size, which is estimated to be that which will reach the minimum economic felling size in half the conversion rotation.

(3) Estimate the average volume per acre which cutting all merchantable trees of economic size (or such greater size as shall have been agreed with the exploiter) will produce, i.e. the permissible cut.

(4) Estimate the average volume per acre which the residual trees now below the minimum felling size adopted will have attained in half the conversion rotation, making suitable allowances for felling damage and mortality in the various size-classes of residual trees.

(5) Divide the conversion rotation into two felling-cycles in the proportion which the volume estimated in (3) above bears to that estimated in (4), i.e.:

$$\text{Number of years in 1st F.C.} = \frac{\text{Number of years in Conversion Rotation}}{1 + \dfrac{\text{Vol. of residual stand per acre after } \frac{1}{2} \text{R.}}{\text{Vol. of permissible cut per acre now}}}$$

Number of years in 2nd F.C. = No. of yrs. in Rotation—No. of yrs. in 1st F.C.

(6) The sustained annual yield in the whole conversion rotation will be

$$\frac{\text{Present Volume of permissible cut in whole forest now} - \text{Safety Reserve, per cent}}{\text{Number of years in 1st F.C.}}$$

and will be harvested by complete exploitation of enumerated trees of the minimum felling size adopted and over, in annual coupes of sufficient area to provide it. The area of the coupes will vary from year to year:

(a) during the first felling-cycle according to the variation in stocking of merchantable trees of exploitable size,

$$\text{Average area} = \frac{\text{Area of whole forest} - \text{Safety Reserve, per cent}}{\text{Number of years in 1st F.C.}}$$

(b) during the second felling cycle according to the variation in stocking of the residual merchantable trees and the time elapsing between the first and second cuttings in the same area. This time will average half the conversion rotation, but will vary from the number of years in the first cycle for the first coupe to the number of years in the second cycle for the last coupe.

$$\text{Average Area} = \frac{\text{Area of whole forest} - \text{Safety Reserve, per cent}}{\text{No. of Years in 2nd F.C.}}$$

A useful check on the annual yield calculated is to multiply the volume of the present permissible cut per acre by the average area of coupes in the first cycle, and the estimated volume per acre of the residual stand after half the rotation by the average area of coupes in the second cycle. Both results should equal the prescribed annual yield. If the safety reserve deduction is made in calculating the coupe area, as shown, this area must, of course, be multiplied by the unreduced volume per acre.

EXAMPLE I

The numbers of trees in the various size-classes are those of the members of the Dipterocarp family quoted by J. Wyatt-Smith from the Nukit Lagong Sample Plot in an unpublished paper entitled *An Ecological Study of the Structure of Tropical Lowland Evergreen Rain Forest in Malaya* (Imperial Forestry Institute, Oxford, 1948). The Malayan Volume Table for Dipterocarp species (E. J. Strugnell, *Malayan Forester*, Vol. 10, p. 97, 1941) has been applied to these figures with a deduction of ten per cent for faulty logs. A safety reserve of ten per cent has also been selected. A rotation of eighty years, a minimum economic diameter limit of 28 inches for exploitation, and a mean growth rate of 12 inches diameter in forty years have been adopted as probably appropriate for the species. The combined felling damage and mortality rates used are, it is hoped, merely pessimistic suggestions.

Present Growing Stock			Mortality percentage	Residual Growing Stock after forty years		
Diam.-class, in.	Number of trees per acre	Volume cu. ft. under bark		Diam.-class, in.	Number of trees per acre	Volume cu. ft. under bark
16–19	2·4	not	40%	28–31	1·4	315
20–23	1·2	exploit-	33%	32–35	0·8	245
24–27	0·8	able	25%	36–39	0·6	240
28–31	1·0	225				
32–35	0·2	62				
36–39	0·2	80	exploited			
40–	2·4	1,123				
		1,490				800

$$\text{1st F.C.} = \frac{80}{1 + \dfrac{800}{1,490}} = 52 \text{ years, and 2nd F.C.} = 28 \text{ years}$$

Per 1,000 acres

$$\text{Annual Yield} = \frac{1,490,000 - 10 \text{ per cent}}{52} = 25,700 \text{ cubic feet under bark}$$

$$\text{Average Annual Coupe in 1st F.C.} = \frac{1,000 - 10 \text{ per cent}}{52} = 17 \cdot 3 \text{ acres}$$

$$\text{Average Annual Coupe in 2nd F.C.} = \frac{1,000 - 10 \text{ per cent}}{28} = 32 \cdot 14 \text{ acres}$$

If present Growing Stock larger

The trees in the 40 inches and over class averaged 50 inches diameter, but the volume shown is that at 40 inches less ten per cent as the volume table goes no higher. If they could provide an extra 210 cubic feet, making 1,700 cubic feet per acre, then:

1st F.C. = 54 years and 2nd F.C. = 26 years,
Annual Yield = 28,000 cubic feet per 1,000 acres, and
Av. Annual Coupe in 1st cycle = 16·6 acres, in 2nd cycle = 34·6 acres.

If the Minimum Exploitable Diameter were raised for the First Cycle

If it were decided that it was economic to raise the minimum diameter to 32 inches in the first cycle the exploitable volume per acre would fall to 1,265 cubic feet. Allowing twenty-five per cent mortality there would be 0·75 trees 40 inches diameter and over per acre in the second cycle, bringing the volume per acre to 1,190 cubic feet. The first cycle would then be forty-one years with an average coupe of 21·9 acres per 1,000 acres, and the second cycle thirty-nine years with an average coupe of 23 acres. The annual yield throughout the conversion period would be 27,700 cubic feet, showing that the estimated increment exceeds the mortality expected. (The volume of each tree of the largest class is taken as that at 40 inches diameter.)

EXAMPLE 2

The numbers per acre of a group of merchantable species (principally *Entandrophragma* spp.) have been calculated from data kindly supplied by A. Foggie about the Assin-Attandaso Utilization Circle in the Gold Coast, in private correspondence in 1948. The minimum girth for exploitation of the species in this circle is 9 feet, the provisional rotation is 100 years, and the mean girth increment is estimated at 1 foot in ten years, but has been calculated below as 4 feet in fifty years for safety. No volume table is available from the Gold Coast, so the yield has been expressed in true square feet of basal area. A safety margin of fifteen per cent has been used in this circle and is retained. The mortality percentages are again pessimistic guesses, smaller than in the last example because of the fewer trees. The output is low for commercial working, but the prescribed yield of this group of species from the circle is at present 0·25 trees per acre, and it is understood that their high local value attracts exploiters who then cut other species as well, which they would not touch without the sprinkling of 'plums'.

233

Present Growing Stock			Mortality percentage	Residual Growing Stock after fifty years		
Girth-class, ft.	Number of trees per acre	Basal area, sq. ft.		Girth-class, ft.	Number of trees per acre	Basal area, sq. ft.
3– 5	0·52	not	25%	7– 9	0·39	?
5– 7	0·25	exploit-	20%	9–11	0·20	1·6
7– 9	0·16	able	12½%	11–	0·12	1·37
9–11	0·24	1·92	exploited			
11–	0·28	3·20				
		5·12				2·97

$$\text{1st F.C.} = \frac{100}{1 + \dfrac{2 \cdot 97}{5 \cdot 12}} = 63 \text{ years, and 2nd F.C.} = 37 \text{ years.}$$

Per 1,000 acres

$$\text{Annual Yield} = \frac{5{,}120 - 15 \text{ per cent}}{63} = 69 \text{ square ft Basal Area.}$$

$$\text{Average Annual Coupe in 1st F.C.} = \frac{1{,}000 - 15 \text{ per cent}}{63} = 13 \cdot 5 \text{ acres.}$$

$$\text{Average Annual Coupe in 2nd F.C.} = \frac{1{,}000 - 15 \text{ per cent}}{37} = 23 \cdot 0 \text{ acres.}$$

INDEX

238

Milton Keynes UK
Ingram Content Group UK Ltd.
UKHW031137141024
449569UK00006B/129